T0271439

Mathematical Models and Environmental Change

This book demonstrates how mathematical models constructed in system dynamics modelling platforms, such as Vensim, can be used for long-term management of environmental change.

It is divided into two sections, with the first dedicated to theory, where the theory of co-evolutionary modelling and its use in the system dynamics model platform is developed. The book takes readers through the steps in the modelling process, different validation tools applicable to these types of models and different growth specifications, as well as how to curve fit using numerical methods in Vensim. Section 2 comprises a collection of applied case studies, including fisheries, game theory and wildlife management. The book concludes with lessons from the use of co-evolutionary models for long-term natural resource management.

The book will be of great interest to students and scholars of environmental economics, natural resource management, system dynamics, ecological modelling and bioeconomics.

Douglas J. Crookes is Associate Professor of Environmental Governance in the School for Public Leadership at Stellenbosch University, South Africa. He has over 22 years of experience in conducting problem-focused and applied research at the interface between the environment and economics.

Routledge Focus on Environment and Sustainability

For more information about this series, please visit: www.routledge.com
/Routledge-Focus-on-Environment-and-Sustainability/book-series/RFES

Mathematical Models and Environmental Change
Case Studies in Long Term Management

Douglas J. Crookes

LONDON AND NEW YORK

from Routledge

First published 2022
by Routledge
2 Park Square, Milton Park, Abingdon, Oxon OX14 4RN

and by Routledge
605 Third Avenue, New York, NY 10158

Routledge is an imprint of the Taylor & Francis Group, an informa business

British Library Cataloguing-in-Publication Data
A catalogue record for this book is available from the British Library

Library of Congress Cataloging-in-Publication Data
Names: Crookes, Douglas J., author.
Title: Mathematical models and environmental change: case studies in long term management/Douglas J. Crookes.
Description: New York: Routledge, 2022. |
Series: Routledge Focus on Environment and Sustainability |
Includes bibliographical references and index. |
Identifiers: LCCN 2021041305 (print) | LCCN 2021041306 (ebook) |
ISBN 9781032163055 (Hardback) | ISBN 9781032163079 (Paperback) |
ISBN 9781003247982 (eBook)
Subjects: LCSH: Global environmental change–Mathematical models.
Classification: LCC GE149 .C76 2022 (print) | LCC GE149 (ebook) |
DDC 333.72068–dc23/eng/20211014
LC record available at https://lccn.loc.gov/2021041305
LC ebook record available at https://lccn.loc.gov/2021041306

ISBN: 978-1-032-16305-5 (hbk)
ISBN: 978-1-032-16307-9 (pbk)
ISBN: 978-1-003-24798-2 (ebk)

DOI: 10.4324/9781003247982

Typeset in Times New Roman
by Deanta Global Publishing Services, Chennai, India

Contents

Acknowledgements

This book would not have been possible without the help of a number of people. I would like to thank the various people who made contributions. Firstly, I thank E.J. Milner Gulland for allowing me to use portions of our work on bioeconomic models that we developed together while I was based at Imperial College London. This formed the basis of Chapter 2. Other people have commented on various chapters and sections: James Blignaut (Chapters 3 and 6); Yaman Barlas and Carryn De Moor (Chapter 5); and Mandy Lombard, Lorien Pichegru, Lynn Shannon, Kelly Ortega and Jai Clifford-Holmes (Chapter 7). I thank my family for their support and for allowing me time to work on this publication. I thank Jon Conrad and Jean-Didier Opsomer. Their 1994 publication on the northern anchovy fishery initiated an interest in co-evolutionary models that has spanned 25 years. This book is the result. I am also thankful to Martin de Wit, who introduced me to system dynamics models and thereby provided the second piece in the puzzle. Finally, but not least, I am also grateful to Ventana Systems UK, specifically Andrew Hill, for training me in numerical methods for optimisation using Vensim. Without him, Chapter 4 would not have been possible.

1   Introduction

The aim of this book is to show how mathematical models constructed in system dynamics modelling platforms, such as Vensim, may be used for long-term management of environmental change. Environmental change is a change or disturbance in the natural environment, for example, through human disturbance or natural processes (Johnson et al. 1997). The concept denotes some form of change in the environment over time. When mathematically modelling these interacting processes, we are referring to non-linear simultaneous differential equations. In biological terms, these processes are referred to as co-evolution, whereas in the fisheries economics literature they are referred to as bioeconomic models. We will utilise the biological term in this book, since it is slightly more encompassing, but many of the models discussed here would fit into the bioeconomic category.

Models for the management of environmental change are ubiquitous. Following Nielsen et al. (2018), there are three types of models: those that provide short-term (tactical) advice, models for the medium term (management strategy evaluation, MSE) and models that provide long-term (strategic) advice. Although opinions vary over the exact time frames for each of these models, Shin et al. (2005) state that tactical models provide advice that is annual, and strategic models provide advice over a period of five to ten years. Therefore, it is concluded that MSE provides advice over a period of one to five years.

Although different models have different purposes and some are useful (Box and Draper 1987), here we are interested in models that provide strategic (long-term) advice. Which models would fall into this category? Although we are not concerned here with a comprehensive review of strategic models, Crookes and Blignaut (2016) found that results of a co-evolutionary model for steel were comparable to artificial neural networks (ANNs) over a ten-year forecast period, after which they digressed. This suggests that co-evolutionary models can be used for strategic advice (but we will further assess this claim during the course of this book).

DOI: 10.4324/9781003247982-1

The aim of this book is to show how co-evolutionary models may be used for long-term management of environmental change. What is co-evolution? Co-evolution occurs when two or more entities interact with each other over time (Conrad 1999). In ecology, this could be two or more species that interact with each other. In this case, these interactions often give rise to the well-known Lotka–Volterra system of equations: predator–prey (Lotka 1925; Volterra 1926, 1928, 1931), interspecific competition (Gause 1932; Gause and Witt 1935), mutualism (May 1982) and parasite–host interactions (Anderson and May 1982). In socioeconomic systems, one entity could be an economic agent (such as a hunter, a poacher or a fisher), and the other could be a biological entity (such as a fish stock). In this case, the equations give rise to so-called fisheries economics models, or (mathematical) bioeconomic models, as we alluded to previously (Clark 1990). Finally, interactions could be between two economic entities, giving rise to inter-sectoral dynamics (Crookes and Blignaut 2016) or interactions between different macroeconomic variables (such as the Goodwin model; Barbosa-Filho and Taylor 2006).

A major constraint in the development of these models is that the biological and economic parameters are often unknown (Butterworth et al. 2010; Nielsen et al. 2018). Various methods have been proposed to overcome this: The non-linear least squares method is provided in packages such as EViews and MATLAB® (Gatabazi et al. 2019). Linear programming is also employed based on the criterion of minimisation of the mean absolute percentage error (MAPE) (Wu et al. 2012). In this book, curve fitting (in other words, the process of deriving a graphical line or mathematical function that best matches a set of data observations) is employed using numerical methods (Markov chain Monte Carlo simulation) in the Vensim modelling platform (Eberlein and Peterson 1992). This method has the advantage of optimising over a wide range of unknown parameter values (Banerjee et al. 2014). The algorithm is a differential evolution/Markov chain hybrid method (Vrugt et al. 2008). Differential evolution belongs to the class of genetic algorithms. It is also possible to use the algorithm for Bayesian inference by specifying a prior distribution and likelihood function (Ter Braak 2006).

Wilensky and Reisman (2006) argue that when curve fitting is employed, it is important that these types of models are validated. The system dynamics modelling literature provides a means by which these types of models may be validated. Furthermore, co-evolutionary models are well suited for use in packages (Swart 1990). The approach used here is, therefore, to build and validate the models using the system dynamics simulation methodology (Sterman 2000).

Once the models have been built and validated, they can be employed for a variety of purposes. There are at least five uses of the models: (1) They can be used to inform the value of biological and economic parameters. This may be an end in itself. These parameters may be unknown in the literature, and curve fitting may be used to estimate these unknown values. (2) The models may contribute towards a better understanding of the behaviour of the system. The choice of model that provides the best fit of the historical data could provide information of the nature of the system and the behaviour of the entities within that system. (3) The models may be used to forecast future values of the entities in question. (4) Most of these model uses are in the context of exploitation. However, co-evolutionary models may also be used as input into game theoretic systems in order to identify conditions for co-operation. (5) Finally, these models may be used for scenario analysis. It involves interrogating the model, for example, through what-if type analysis. All of these uses (points 1–5) suggest that these types of models, when properly built and validated, could contribute towards providing advice for the management of natural resources and other economic and financial systems.

The structure of the book is as follows: In the next chapter, we provide a review of different models of exploitation. Then, we discuss the simulation modelling technique, highlighting steps in the modelling approach and validation. In Chapter 4 we elaborate on the method of curve fitting, showing how to implement numerical simulation methods in Vensim. We highlight a number of different applications. One is forecasting populations. An application to rhino management is considered in Chapter 5. This technique emphasises the use of different models and comparing the dynamics thereof to determine which behaviour is most realistic.

Next, the co-evolutionary models are used in a modified prisoner's dilemma game in order to determine under what conditions co-operation is possible. This is the topic of Chapter 6. Chapter 7 provides an example of these co-evolutionary models used for providing tactical management advice, using the case of the African penguin. This example highlights how incorporating environmental stochasticity may enhance the realism of these models. Finally, Chapter 8 concludes with a discussion of the findings.

References

Anderson, R.M. and May, R.M., 1982. Coevolution of hosts and parasites. *Parasitology*, 85(Pt 2), pp.411–426.

Banerjee, S., Carlin, B.P., and Gelfand, A.E., 2014. *Hierarchical Modeling and Analysis for Spatial Data* (2nd edn). London, UK: Chapman & Hall.

Barbosa-Filho, N.H. and Taylor, L., 2006. Distributive and demand cycles in the US economy—A structuralist Goodwin model. *Metroeconomica*, 57(3), pp.389–411.

Box, G.E.P. and Draper, N.R., 1987. *Empirical Model-Building and Response Surfaces*. New York: John Wiley & Sons.

Butterworth, D.S., Bentley, N., De Oliveira, J.A., Donovan, G.P., Kell, L.T., Parma, A.M., Punt, A.E., Sainsbury, K.J., Smith, A.D., and Stokes, T.K., 2010. Purported flaws in management strategy evaluation: Basic problems or misinterpretations? *ICES Journal of Marine Science*, 67(3), pp.567–574.

Clark, C.W., 1990. *Mathematical Bioeconomics: The Optimal Management of Renewable Resources* (2nd edn). New York: Wiley.

Conrad, J.M., 1999. *Resource Economics*. New York: Cambridge University Press.

Crookes, D.J. and Blignaut, J.N., 2016. Predator-prey analysis using system dynamics: An application to the steel industry. *South African Journal of Economic and Management Sciences*, 19(5), pp.733–746.

Eberlein, R.L. and Peterson, D.W., 1992. Understanding models with Vensim™. *European Journal of Operational Research*, 59(1), pp.216–219.

Gatabazi, P., Mba, J.C., Pindza, E., and Labuschagne, C., 2019. Grey Lotka–Volterra models with application to cryptocurrencies adoption. *Chaos, Solitons & Fractals*, 122, pp.47–57.

Gause, G.F. 1932. Experimental studies on the struggle for existence. *Journal of Experimental Biology*, 9(4), pp. 389–402.

Gause, G. and Witt, A., 1935. Behavior of mixed populations and the problem of natural selection. *The American Naturalist*, 69(725), pp.596–609.

Johnson, D.L., Ambrose, S.H., Bassett, T.J., Bowen, M.L., Crummey, D.E., Isaacson, J.S., Johnson, D.N., Lamb, P., Saul, M., and Winter-Nelson, A.E., 1997. Meanings of environmental terms. *Journal of Environmental Quality*, 21(5), 581–589. doi:10.2134/jeq1997.00472425002600030002x

Lotka, A.J., 1925. *Elements of Physical Biology*. Baltimore, MD: Williams and Wilkins.

May, R.M., 1982. Mutualistic interactions among species. *Nature*, 296(5860), pp.803–804.

Nielsen, J.R., Thunberg, E., Holland, D.S., Schmidt, J.O., Fulton, E.A., Bastardie, F., Punt, A.E., Allen, I., Bartelings, H., Bertignac, M., and Bethke, E., 2018. Integrated ecological–economic fisheries models—Evaluation, review and challenges for implementation. *Fish and Fisheries*, 19(1), pp.1–29.

Shin, Y.J., Rochet, M.J., Jennings, S., Field, J.G., and Gislason, H., 2005. Using size-based indicators to evaluate the ecosystem effects of fishing. *ICES Journal of Marine Science*, 62(3), pp.384–396.

Sterman, J.D., 2000. *Business Dynamics. Systems Thinking and Modelling for a Complex World*. Boston, MA: Irwin McGraw-Hill.

Swart, J., 1990. A system dynamics approach to predator-prey modeling. *System Dynamics Review*, 6(1), pp.94–99.

Ter Braak, C.J., 2006. A Markov Chain Monte Carlo version of the genetic algorithm Differential Evolution: Easy Bayesian computing for real parameter spaces. *Statistics and Computing*, 16(3), pp.239–249.

Volterra V., 1926. Fluctuations in the abundance of a species considered mathematically. *Nature*, 118(2972), pp. 558–560.

Volterra, V., 1928. Variations and fluctuations of the number of individuals in animal species living together. *Journal du Conseil Permanent International pour l'Exploration de la Mer*, 3(1), pp.3–51.

Volterra, V., 1931. *Lectures on the Mathematical Theory of Struggle for Life*. Paris: Gauthier-Villars.

Vrugt, J.A., Hyman, J.M., Robinson, B.A., Higdon, D., Ter Braak, C.J., and Diks, C.G., 2008. *Accelerating Markov Chain Monte Carlo Simulation by Differential Evolution with Self-Adaptive Randomized Subspace Sampling (No. LA-UR-08-07126; LA-UR-08-7126)*. Los Alamos, NM (United States): Los Alamos National Lab (LANL).

Wilensky, U. and Reisman, K., 2006. Thinking like a wolf, a sheep, or a firefly: Learning biology through constructing and testing computational theories—An embodied modeling approach. *Cognition and Instruction*, 24(2), pp.171–209.

Wu, L., Liu, S., and Wang, Y., 2012. Grey Lotka–Volterra model and its application. *Technological Forecasting and Social Change*, 79(9), pp.1720–1730.

2 Models of co-evolution

2.1 Traditional bioeconomic models

Under this section, we briefly describe three types of traditional bioeconomic models: (1) static models, (2) dynamic models under open-access exploitation and (3) dynamic models under the profit-maximising behaviour of a resource owner.

2.1.1 Static models

The basic bioeconomic theory follows the pioneering work of Gordon (1954), Schaefer (1954, 1957) and Smith (1968, 1969). Biological population growth in continuous time may be written as

$$\frac{dx}{dt} = F(x)$$

$F(x)$ in the standard Verhulst logistic function takes the form $F(x) = rx(1 - x/K)$, where r is the intrinsic growth rate, x is the stock density and K is the carrying capacity. This formulation for density-dependent growth assumes that the maximum sustainable yield occurs at $0.5K$ and yields the traditional symmetrical inverted U shape in population ecology.

For long-lived species such as mammals, a growth function with a maximum sustainable yield at $0.5K$ may not be an adequate approximation of species production. For example, empirical data suggest that the maximum sustainable yield (MSY) of hunted populations should approximate 80% of carrying capacity for long-lived species such as elephants and fin whales, $0.5K$ for species such as deer and mice, and less than $0.5K$ for species such as insects (Fowler 1981, 1984; Milner-Gulland and Mace 1998). It would therefore be preferable to allow for increased flexibility in the formulation of the growth model, so that maximum sustainable yields may occur outside

DOI: 10.4324/9781003247982-2

of 0.5K. Several adaptations to the original theory have been made to allow for this variability. Fox (1970) develops a function that is skewed to the left of 0.5K, based on the Gompertz growth specification. Pella and Tomlinson (1969) produce a more general framework by introducing a power relationship into the logistic model (the exponent m in Equation 2.1). It may be shown that, for specific parameter values, the Fox model is a special case of the more general Pella and Tomlinson model (Crookes 1997). Here we consider a simplified form of the Pella and Tomlinson model, consistent with our previous discussions:

$$F(x) = rx\left(1 - \left(\frac{x}{k}\right)^m\right), \quad m > 0 \tag{2.1}$$

The parameter m determines the skewness of the function. If $m = 1$, the curve collapses to the logistic function with the symmetrical distribution. If $m > 1$, the curve is skewed to the right, while for values $0 < m < 1$ the function is skewed to the left.[1] Assuming adequate data is available, the correct functional form of the model (based on the parameter m) may either be derived through the use of standard regression and other statistical techniques or based on literature estimates.

Equation 2.1 describes the growth in the absence of any hunting activity or any other predation. Under this formulation, populations will increase and stabilise at the carrying capacity. However, for our purposes it is preferable to include a harvest function h, usually written $h = qEx$ (where q is the catchability coefficient and E is the hunting effort). For a given level of harvest, populations will stabilise when the growth function $F(x) = h$. For the simplest hunting model, therefore, an increase in hunting h will increase the slope of $h = f'x$ (since effort is increasing), and stock levels in equilibrium will decline (f' increases from f_0 to f_1, x falls from x_0 to x_1; see Figure 2.1).

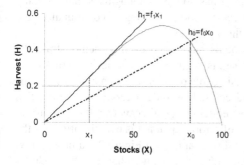

Figure 2.1 Effect of an increase in harvests on stocks in equilibrium ($m = 4$).

These types of models are widely used in the fisheries economics literature (e.g. Opsomer and Conrad 1994) and have also seen some application in the terrestrial resource management literature (e.g. Milner-Gulland and Leader-Williams 1992).

A substantial literature has grown out of the relationships described here between harvests and stocks, and we will elaborate on much of this work next. Two brief remarks may, however, be made at this stage. Firstly, it is clear that the effects of various policy instruments will ultimately work through the coefficient f in the harvest function. This is, however, a gross simplification since the relationships shown here are equilibrium conditions and therefore conceal the underlying short-term dynamic effects on harvesting cost structures and prices that characterise a convergence to equilibrium. Secondly, it is clear that policies that increase the slope of the function f may have the effect of increasing harvests, in the short term and/or in the long term, depending on the exact location of the equilibrium points before and after the policy intervention. This outcome is due to the recovery of stocks following a reduction in effort. The actual processes involved in reaching an equilibrium solution following a policy intervention are elaborated upon in a subsequent section.

2.1.2 *Institutional aspects and bioeconomic modelling*

Differing ownership regimes affect the behaviour of harvesters, and subsequently the equilibrium levels of stocks that are achieved. Three types of ownership regimes may be distinguished: (1) pure "open access" harvesting, where both entry to a natural resource and harvesting is unrestricted; (2) regulated open access, where access to the resource is unrestricted, but where harvest levels are restricted; and (3) private property, where entry and harvests are usually restricted. The idea that all resources in the hands of communities were essentially open access led to the concept of the "tragedy of the commons". However, it has now been recognised that, in many instances, rules exist for the access to and harvesting of communally held resources (e.g. MacKay and Acheson 1987).

Traditional bioeconomic models have been used to demonstrate the impacts of different property rights regimes on equilibrium stocks. As indicated earlier, profit-maximising sole owners seek to maximise the difference between revenues and costs. In standard economic theory, this occurs when marginal revenues equal marginal costs (i.e. the increase in revenues for an additional unit of effort expended is exactly offset by the additional costs of the activity). For example, in Figure 2.1 a profit-maximising stock level may be characterised by x_0, with harvests at h_0. Under this property rights regime, bioeconomic theory based on the traditional logistic model predicts that (1) harvest levels will always be at stock levels greater than the MSY. The economic objective of profit maximisation is therefore a more conservative policy than the biological objective of maximum sustainable yield. Only where total harvest costs are zero will stock levels drop to levels equal to MSY. (2) Extinction is not possible under the profit-maximising conditions of a sole owner. This result, however, only holds in the static case where the time horizon of the resource owner is myopic.

These results assume that the owner can keep other entrants from harvesting the resource. Where entry is not restricted, economic rents will be dissipated and the equilibrium will tend towards the open-access solution. Under open-access harvesting, the logistic bioeconomic model predicts that (1) the open-access equilibrium is always at stock levels below the profit maximisation levels (e.g. at point x_1 in Figure 2.1). (2) Contrary to what is sometimes supposed, this does not necessarily result in the extinction of resources. (3) On the other hand, if unit costs are high enough, open access in the logistic model will produce stock levels below or equal to the MSY, where it has been demonstrated that equilibrium stock levels are less stable and more vulnerable to perturbations than at levels above the MSY. (4) Since the owner can restrict entry, the profit-maximising price is higher than would be possible under competitive (open access) conditions. Overall demand is, therefore, higher for an open-access resource. Price adjustments are, therefore, important policy tools for shifting bioeconomic equilibria towards sustainable levels.

What about the situation where "rules" govern access and harvesting of a resource? In terms of the bioeconomic model previously described, stock levels under communal management would occur between the pure open access and private property levels, depending on the degree to which entry and harvests are limited. However, the harvesting outcomes predicted by the models do not depend primarily on the ownership regimes in place, but rather on the abilities of the resource 'owners' to enforce and maintain the rules governing the harvesting of resources.

2.1.3 Dynamic open-access behaviour

The previous discussion focussed on static, equilibrium levels of stocks and harvesting. Dynamic entry–exit behaviour in open-access fisheries is an important aspect of traditional bioeconomic models, since it becomes possible to investigate the time path of stocks towards equilibrium. These types of models have seen numerous applications in the marine resource literature (see for example Wilen 1976; Opsomer and Conrad 1994; Bjorndal and Conrad 1987; Conrad 1995). It is commonly assumed that the dynamic interactions of stocks and harvesting efforts are based on the Lotka–Volterra predator–prey system, where harvesters will enter an open-access resource in the pursuit of profits. Where positive rents may be generated, harvesting efforts increase, and where losses are made, efforts decrease. A simple expression of this relationship may be given as follows:

$$\frac{dE}{dt} = n(pqx - c)E \tag{2.2}$$

where the term in brackets is net revenue (priceless unit costs, assumed constant) for a given level of effort E, and n is an adjustment parameter indicating the response of effort (E) to changes in net profits. Combining this relationship with the stock growth under harvesting:

$$\frac{dx}{dt} = F(x) - qEx \tag{2.3}$$

gives the dynamic interactions between harvesting effort and stock densities. It is evident from these two relationships that the dynamics of effort over time (Equation 2.2) is dependent on stocks (x), and the dynamics of stocks over time (Equation 2.3) is dependent on effort (E). The interrelatedness of harvesting effort and stock densities, therefore, determine the overall transition of the system over time. It is then possible to investigate the short-term response of stocks and effort to changes in parameter values, in order to determine the conditions under which convergence (or lack of convergence) to equilibrium is achieved (Figure 2.2). Figure 2.2A (top graph) indicates a corner solution, where capital enters the fishery and the prey decreases to zero. Figure 2.2B (middle graph) indicates a limit cycle, where predator and prey oscillate over time, never reaching a stable equilibrium. In this system, equilibrium is also a possible outcome (Figure 2.2C, bottom graph), with predator and prey populations converging on a stable value. It is also possible for a second corner solution to occur, where the predator decreases to zero and the prey increases to carrying capacity (in the figures, prey is expressed as a proportion of carrying capacity, so in this case $E = 0$ and $x/k = 1$).

These models may also be generalised to include opportunity costs and other dynamic adjustment behaviour. Although historically open-access entry–exit models have primarily been in the context of fisheries, Bulte and Horan (2001) use this framework to consider decisions to convert from hunting to agricultural production. In their model, the dynamics of effort is based on the profitability of agriculture relative to hunting, and stock dynamics are based on time spent hunting and the habitat available for wildlife.

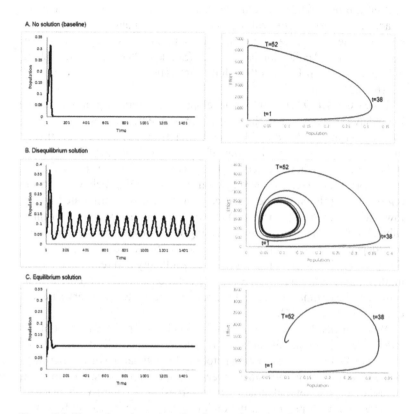

Figure 2.2 Phase plane dynamics of predator and prey species. Notes: See Equations 2.2 and 2.3.

2.1.4 Profit maximisation over time

Thus far it has been assumed that resource users have a myopic time horizon. Often it is the case, however, that the goal of the resource owner is to maximise the discounted present value of revenues (or profits) over an extended time horizon. This situation may occur when an owner has an interest in the sustainable use of a resource, for example (1) wildlife authorities maximising net returns to tourism subject to the availability of stocks; (2) private game ranches where trophy hunting is restricted to recreational use; or (3) in cases where private harvesting rights are in the hands of local communities, and outsiders are excluded. However, it is not necessary for the resource to be under the sole ownership of the resource users for such conditions to apply. On the contrary, non-owners who have a stake in the resource may act as *de facto* stewards of the resource, effectively excluding outsiders, provided there is sufficient cause or justification for such actions. Examples where collective rules governing behaviour and excluding outsiders have emerged spontaneously, in spite of users not owning the resource, include driftwood collection in Yorkshire (Sugden 1989) and lobster fishing in Maine (Acheson 1987), although the motivations for such behaviour are poorly understood.

It was Clark (1973) who first developed the theory of profit-maximising behaviour over time in the case of renewable natural resources. More recently, Milner-Gulland and Leader-Williams (1992) applied this theory in the case of the illegal hunting of ivory and rhino horn in Luangwa Valley, Zambia, where dealers seeking economic rents in the ivory and rhino horn trade exerted a strong influence on hunting behaviour in the region. The problem for the resource owner is usually formulated as follows: to select harvests (or effort) that maximises net revenues over time, subject to the dynamics of the underlying stocks. Harvest rates will therefore depend on the existing levels of stocks. Such a dynamic optimisation problem is solved using optimal control theory and requires complex mathematical techniques to solve. The solution to such a problem indicates that the optimal strategy of the resource harvester is to harvest at the optimal stock level (x^*).

When stocks deviate from x^*, the optimal strategy is to harvest at the rate that drives stocks towards x^* as quickly as possible. Therefore, for initial stocks less than the optimal stocks, the optimal harvest policy is to abstain from hunting until stocks reach the optimum. For initial stocks in excess of x^*, the optimal strategy is to harvest at h_{max} until x^* is reached. Harvests will then proceed at the optimum for $x = x^*$. It should be noted, however, that the optimal strategy is not necessarily a sustainable strategy. For example, where the rate of return on the biological populations is less than the discount rate, it may be optimal to "disinvest" (i.e. to harvest stocks to extinction). This is explained as follows: the discount rate represents the opportunity cost of capital or the return that may be achieved from alternative investments of capital. When the returns to the natural resource are less than the opportunity cost of capital, it may be optimal to cash in on the stocks as soon as possible and invest the proceeds in the higher-yielding alternative investments.

2.2 Recent advances in bioeconomic modelling

The basic framework of bioeconomic theory is constantly being extended to handle complexities encountered in natural systems. In this section we briefly consider three recent additions: (1) multispecies models, (2) spatially explicit models and (3) stochastic models. Many of these applications have been in the fisheries literature, although some are beginning to find their way into the terrestrial resource management literature as well.

2.2.1 Multispecies models

Two types of multispecies interactions are of particular concern to us.[2] Interactions may occur between species with regard to competition for resources (food, habitat) or may be characterised by predator–prey behaviour. Under density dependence, where species interact in one of these ways, a decrease in the density of one species, for example through hunting, may result in either an increase or decrease in the densities of other species. In the case where species are competing for resources, hunting of one species may reduce the density of that species in one of two ways: (1) directly, through reducing the stocks of hunted species relative to unhunted species; or (2) indirectly, by increasing the resources available to unhunted species. This effectively increases the carrying capacity of the unhunted species, enabling greater growth and effectively inhibiting the recovery of hunted species. Therefore, instead of growth rates of hunted species increasing at low densities, low growth rates may persist and even drive stocks to extinction. A second case where hunting may increase the vulnerability of stocks is when a facultative relationship with regard to food resources exists. An example is forest duikers, which are dependent on primates and other canopy dwellers for the dislodging of fruits. Hunting of these food providers may increase competition for food amongst non-target species, thereby reducing numbers.

Although many of these interactions are complex and difficult to predict, bioeconomic models have been used to understand some of the mechanisms and predict critical harvest levels. Relatively simple adjustments to the standard density-dependent growth formulation may be made to account for species interactions. In a two-species system, for example, a typical continuous time formulation could be

$$\frac{dx}{dt} = F_1(x) + axy - h_1$$

$$\frac{dy}{dt} = F_2(y) + bxy - h_2$$

where x and y are species with unique growth characteristics F_1 and F_2, and subject to independent harvesting h_1 and h_2. The variables axy and bxy determine the nature and extent of species interactions. For $a < 0$ and $b < 0$, both species are competing for resources. When $a > 0$ and $b < 0$, a predator–prey

model emerges where species *x* feeds on species *y*. A number of other specifications are possible. Even these simple specifications can provide quite
useful insights into biological interactions. For example, in the case where
duikers (*x*) are reliant on primates (*y*), and not vice versa, we have a situation where $a > 0$ and $b = 0$. The model is easily generalised to include further
species interactions. For example, suppose that species *z* is now included in
the system of equations. It is then possible to model interactions between,
say, species *x* and *z*, and not between *y*. In this case, the growth function for
species *x* would contain an additional *cxz* term, and the growth function for
z would contain a *dxz* term. Obviously, the challenge in these models is to
derive appropriate estimates for the biological parameters *a*, *b*, *c* and *d*.

Several important insights may be derived from the modelling of such
interactions. For example, studies have indicated that when two species of
varying densities are hunted, while it may not be economically viable to only
hunt the species of low density, the economic viability of hunting higher-
density species may drive stocks of these low-density species to extinction.
Clearly then, to ignore competitive interactions, where these exist, could
have adverse consequences for the management of populations. In addition,
equilibria in multispecies systems may be less stable and more vulnerable
to changes in harvesting levels. For example, modelling results indicate that
when competitive interaction occurs between species, stocks may be driven
to extinction at harvest, and stock levels approximate the MSY in the individual species case (Clark 1990). This is a reason offered for the collapse of the
Pacific sardine fishery in the 1940s and 1950s, where harvesting of sardines
precipitated the recovery and ultimate dominance of an anchovy population.
Finally, open-access exploitation is more likely to lead to the elimination of
desirable species than a profit-maximising exploitation rate. In practice, multispecies models have been used *inter alia* to explore interactions between
whale species (May et al. 1979; Clark 1990), and rhinos and elephants (work
cited in Milner-Gulland and Mace 1998). More recently, Damania et al.
(2003) investigated the effects of hunting tigers on herbivore populations.

2.2.2 Spatially explicit models

The modelling of spatial systems is important when natural resource zones
are heterogeneous, and may be useful for analysing spatial management policies such as natural refuges, "no-take" areas and rotating harvest zones. Such
models may also be used to investigate the effects of harvesting behaviour
over time, both as a result of these policies as well as the effects of terrain
(for example accessibility) and distance to hunting area. In instances where
the study of spatial dynamics is of interest, the standard bioeconomic model
previously discussed may be generalised to include spatial characteristics.

The majority of applications of spatial modelling are found in the fisheries literature (e.g. Allen and McGlade 1986, 1987; Sanchirico and Wilen 1999, 2000). More recently, Hofer et al. (2000) developed a spatial profitability model to consider the distribution of net benefits from illegal hunting in different locations in Serengeti, East Africa. In these types of models, a patch is usually defined as a homogeneous area where a given population of a resource is located in time *t*. A patch is usually selected in terms of a Ricardian hypothesis of maximising net revenues, subject to the costs of harvesting from a particular patch. Harvesting costs in this spatial framework may be a function of the distance to a patch, capital investment in harvesting equipment, opportunity costs, and, where harvesting is illegal, the cost of penalties associated with an arrest.

Developing a dynamic system with interactions between hunters and prey species is more difficult, although attempts have been made to include this aspect (e.g. Allen and McGlade 1986; Sanchirico and Wilen 1999). Including stock dynamics allow different patterns of harvesting behaviour to be modelled explicitly and compared, both in terms of effects on stock dynamics as well as the relative profitability of harvesters. For example, the models of Allen and McGlade (1986, 1987) differentiate between two categories of harvesters: (1) a conservative group that only harvests where there is a certainty of finding a resource stock, and (2) a more risk-embracing group that may venture into new territory in search of higher stock densities (and hence, a higher return). Modelling results indicate that the long-term profitability is higher amongst those more willing to embrace risk and seek out new harvesting opportunities. Other aspects that have been included in these more sophisticated models include the quality of the information available regarding stocks and the degree of communication between harvesters.

Also, by defining the nature of stock dispersion between patches, various types of realistic stock behaviour and strategies for management of the resource may be modelled (e.g. Carr and Reed 1993). For example, a distinction may be drawn between a *closed system*, where there is no dispersal between patches; a *fully integrated system*, where dispersal occurs between all patches; and a *source-sink* or *multiple source* framework, where dispersal occurs from certain sources only. In addition, a variety of *spatially linear* configurations are also possible. For example, limited distance dispersal could be characterised by population movements only to adjacent cells.

2.2.3 Stochastic models

Stochastic conditions are usually introduced in models to reflect various types of uncertainty. These include *process* uncertainty, the inherent variability that is found in natural systems; *observation* uncertainty, which indicates data limitations and difficulties in getting information on the system; and *model* uncertainty, which relates to the fit of the model to the actual data.

In practice, techniques used for introducing observation error and process error into models may be very similar. Observation error and changes in the parameters over time (for example, through habitat destruction) may be introduced by including linear trends, carrying out a sensitivity analysis in the base case parameters and/or allowing a wider margin for errors (e.g. by increasing the coefficient of variation in the model). Combinations of these are required when interactions between variables are important (Milner-Gulland et al. 2001).

We will briefly describe two types of stochastic models that have been applied in the natural resource management context. The first is the *continuous time* model where the growth rate is random and *dependent* on the stock level x. Under such conditions, the time derivatives of x are not defined in an ordinary sense. Such stochastic differential equations are complex and require advanced mathematical techniques to solve. Furthermore, solutions to these equations frequently yield higher-order, non-linear, non-homogeneous ordinary differential equations for which an analytical solution is unlikely (Conrad 2004). However, much may be learnt about the dynamics of the system through studying these models. Furthermore, a number of special cases have been documented (based on realistic assumptions) that provide an indication of the steady-state distributions of stocks and harvests over time. A growing number of applications in the literature are based on such approaches (e.g. Beddington and May 1977; May et al. 1979; Allen and McGlade 1986; Hortsthemke and Lefever 1984).

The second type of model that has been used in the fisheries literature is a *discrete time* stochastic model where the growth rate is random and *independent* of the stock level x. The solution to this optimisation problem over time may be found using the method of Lagrange multipliers. These categories of models, although complex, are more tractable than the first category, and unique solutions may be more readily derived. Conrad (1995) uses such a model that yields a result that is basically an adaptation of the golden rule (the optimal harvest rate), to develop an approximately optimal feedback control policy. Managers may use such a policy to determine the optimal levels of effort for a given period (e.g. one year), under stochastic stock conditions. This particular model is useful where short-term fluctuations in the stock are likely to render long-term stock equilibria irrelevant for management purposes. For example, in the model of Conrad, for a high catchability coefficient, the fishery would not be viable over the long term. However, due to temporary increases in abundance, harvesting is profitable in the short term for those same parameter values. In theory, of course, the contrary may also occur, in that long-term equilibria may be profitable, while short-term fluctuations in stocks under sustained harvesting may increase the risks of extinction.

Under stochastic conditions, two policy effects may be predicted. Firstly, under regular stock assessments, a higher level of offtake may be possible in some years than would otherwise be the case. Furthermore, short-term harvesting of stocks may be possible, although harvesting over the long term may be unsustainable. Secondly, under stochasticity, a more conservative harvesting strategy could be pursued. This may be preferred where regular stock assessment is not possible (either for cost or logistical reasons). Clearly, though, this will have a greater effect on the livelihoods of harvesters dependent on the resource. In optimal harvest models under uncertainty, it has also been observed that simple harvesting strategies generally perform better than more complex strategies, the latter potentially being more susceptible to error and bias (Milner-Gulland et al. 2001).

2.3 Opportunity cost models

Optimisation models are also useful where biological populations are undermined by land conversion and other conflicts with development. Such models emerged from early applications of capital theory to natural resources in the context of irreversible development (e.g. Fisher et al. 1972). More recently (e.g. Swallow 1990; Swanson 1994) these have been extended to consider the effects on renewable resources. Usually, given the complexities of the analytical solutions, these models have a two-sector partial equilibrium framework. Swallow (1990), for example, considers the trade-offs between a fisheries sector and a development sector that undermines production through changes in environmental quality. Recent applications to the management of terrestrial species have focussed on interactions between protected areas authorities and agro-pastoralists. These models have been used to investigate not only the opportunity costs of land (Bulte and Horan 2001; Schultz and Skonhoft 1996), but also the opportunity costs of living with wildlife (e.g. through the transmission of diseases, damage to crops; see for example Skonhoft 1998, and Sutton and Jarvis 1998). Such models usually take the form of a problem in optimal control theory, where the objective for the resource owner is to optimise net social revenues for the land uses over an infinite time horizon.

Predator–prey type models have also been used to consider the opportunity cost of land as a cause of species extinctions, combined with metapopulation theory (e.g. MacArthur and Wilson 1967; Levins 1970). In these models, interactions are not defined primarily as those between species, but rather between patches of land dominated by a particular homogeneous land use. For example, one equation may define the number of cells containing hunted species in a given spatial area (for example in a park), and the other equation may define the number of cells occupied by agro-pastoralists. The dynamics of the system would then be limited by the total availability of cells. Examples of such models include May (1994) and Bulte and Horan (2001). Adjustment occurs depending on the parameters of the system and may be used to predict the optimal area needed to be set aside for conservation (prey-only cells) vis-à-vis other land uses. In spite of their limitations (see for example Harrison 1994) these models are useful for considering the dynamics of the system.

2.4 Further functional forms

To recap, the traditional predator–prey model models the interactions between a predator population (for example a fox) and a prey species (for example rabbits). The relationship is given by

$$\frac{dx}{dt} = F(x) - sxy \tag{2.4}$$

$$\frac{dy}{dt} = sxy - my \tag{2.5}$$

where x is the prey species, y is the predator, $F(x)$ is some biological growth function, usually the logistic function, and s and m are positive constants. If $y = 0$ (no predators), it is easily shown that the prey species grows to the environmental carrying capacity k, whereas if $x = 0$ (no prey), then the predator population dies out at an exponential rate. The assumption is that the predator population is the only agent that controls the prey and that the prey is the sole food source of the predator.

2.5 Growth functions

Earlier, we mentioned the basic growth functions. Here we elaborate in more detail the mathematical form of these functions. The first, as we mentioned earlier, is the logistic growth function:

$$F(x) = rx\left(1 - \frac{x}{k}\right) \tag{2.6}$$

where r is the intrinsic growth rate and k is the environmental carrying capacity. Maximum growth occurs at $k/2$. With k finite, and assuming a logistic growth function, the predator–prey system will converge on one nontrivial equilibrium point, $x^* = m/s$ and $y^* = b(1 - m/ks)/s$.

A second function is the Fox (1970) model, based on the Gompertz growth function, and assumes that maximum growth occurs at less than half the maximum stock size:

$$F(x) = rx\left(1 - \frac{\ln x}{\ln k}\right) \tag{2.7}$$

The third is the Pella and Tomlinson (1969) specification:

$$F(x) = rx - \gamma x^z \tag{2.8}$$

where x is the biomass, and γ and z are parametric constants. Setting surplus growth equal to zero gives the maximum population size, equivalent to the environmental carrying capacity k:

$$x_{max} = k = \left(\frac{r}{\gamma}\right)^{\frac{1}{z-1}} \tag{2.9}$$

It may be shown that, by setting $z = 2$, the logistic growth function is obtained, and similarly, as z approaches 1, the growth function approaches the Gompertz exponential form of the Fox model.

The choice of growth function will reflect the underlying theoretical underpinnings of the behaviour of the prey species and the available data. These growth functions have the advantage that they are not data intensive, in comparison with other growth functions. The different functional specifications determine the different rates at which growth occurs. In the next sections, we will consider the bioeconomic theory of fisheries.

2.6 Agent expectations

These models have also been coupled with dynamic econometric models based on how expectations are formed. For example Crookes (1997) models fishing decisions based on the adaptive expectations hypothesis, in that expectations are formed regarding what stocks will be in the future based on their expectations of what current stocks are. Berck and Perloff (1984) contrast myopic and rational expectations within a general dynamic model of an open-access fishery. These types of models indicate that short-term adjustments may differ from long-term steady-state values. These are two examples where individual expectations may be incorporated into predator–prey models.

The Cobb–Douglas production function is not the only harvest function that is possible. In the next section we will consider two other harvest functions, namely Baranov's catch equation and Holling's type III predation.

2.7 Alternative catch functions

2.7.1 Baranov catch equation

Although the Cobb–Douglas production function is the most common in the open-access literature, there are other important harvesting functions in the fisheries literature. One example is the Baranov catch equation. The harvest at the end of the season is

$$h^b = \frac{F}{F+M}\left(1-e^{-(F+M)T}\right)\bar{x}_0 \tag{2.10}$$

where $\bar{x}_0$ is the initial population abundance and F and M are the fishing and natural mortality rates, respectively, and T is usually 1 (Liu and Heino 2014). Liu and Heino (2014) assume a Cobb–Douglas production function for F of the form

$$F_t = ah_t = aqE_t x_t^\beta \tag{2.11}$$

where a is a scaling parameter and β captures the potential of a non-linear response of harvests to changes in abundance. Catch h^b therefore becomes

$$h_t^b = \frac{F_t}{F_t+M}\left(1-e^{-(F_t+M)}\right)\bar{x}_0 \tag{2.12}$$

2.7.2 Holling's type III predation

The Holling's type III predation model is another example of a non-linear harvest function, assuming a convex–concave production function (Bulte 2003). Following Bulte (2003), the harvest function is given as

$$h(x) = \frac{x^2}{A\left(1+\dfrac{1}{\gamma^2}x^2\right)} \tag{2.13}$$

where $A = \dfrac{\gamma^2}{\delta}$. It is important to note that, unlike the Cobb–Douglas production function that is a function of both stocks and effort, Holling's predator function is a function of only stocks. The parameter γ indicates the level of prey abundance at which saturation begins to occur, and δ indicates the maximum level of offtake per period, in other words the saturation level of harvesting. Bulte (2003) states that "these parameters are not without economic meaning but it takes a structural model to shed more light on this issue" (p. 30). In the next section we will consider a number of other issues related to the development of Lotka–Volterra (LV) models.

2.8 Other model issues

2.8.1 Data

Because LV models are relatively simple, the greatest challenge is not the model building itself, but the obtaining of data. For example, if the intention is to build an inter-sectoral predator–prey model, data is needed on prices and costs for each sector, along with growth rates, carrying capacity estimates (maximum production) and rates of adjustment. All bioeconomic models are data intensive; however, an advantage of modelling bioeconomic systems is that often historical time series data on economic phenomena are available, even at a high level of disaggregation. This makes these types of models very useful for applied modelling.

2.8.2 Criticisms

The LV model is not without criticism. A weakness of the predator–prey system is that it is not structurally stable (Clark 1990). However, Clark goes on to argue that it is the instability that adds realism to the system. Cyclical behaviour is common in economic systems, as demonstrated by the ubiquitous use of the Goodwin model. Other advantages of the framework are that it is simple and transparent and the interpretation of the parameters is straightforward (Begon et al. 1996). This has resulted in its widespread use in a variety of applications and settings.

2.9 Non-bioeconomic models

Thus far we have focussed primarily on bioeconomic models. In the case of bioeconomic models, and depending on the use of the model, how the model is specified is very important. For other systems and uses, it is not always important how predators and prey species are defined, or indeed that there is some form of predatory behaviour. For example, Ibáñez et al. (2008) developed a model for a competitively exploited aquifer, where groundwater is the prey and the irrigated hectares are the predators. Crookes (2018) developed an LV model for a water supply system, where the predator is the number of dams built and the prey is the dam storage levels. Dendrinos and Mullally (1981, 1983) developed an urban dynamics model, defining the urban population as the predator and per capita income as the prey. For a similar urban dynamics application, Orishimo (1987) defined population as the prey, and land price as the predator. Puliafito et al. (2008), on the other hand, applied a general approach without specifying either variable as prey or predator, since they argue that they "could probably find different intuitive justification to choose one or the other option".

Notes

1 The relationship between x_{msy}, the population at which maximum sustainable yield is attained, and m the skewness parameter may be represented as follows:

$$x_{msy} = \frac{1}{(m+1)^{\frac{1}{m}}} k, 0 < x_{msy} < k$$

so that if $m = 1$, x_{msy} is at the usual $0.5k$ of the logistic function. Furthermore, from this relationship it is easily seen that if $m = 0.2$, x_{msy} is at approximately $0.4k$ and if $m = 2.4$, x_{msy} is approximately $0.6k$.

2 The cases of parasite–host and symbiotic interactions are therefore ignored.

References

Acheson, J., 1987. The lobster fiefs revisited: Economic and ecological effects of territoriality in Maine lobster fishing. In B. Mackay and J. Acheson (Eds), *The Question of the Commons: The Culture and Ecology of Communal Resources.* Tucson, AZ: University of Arizona Press, pp.37–65.

Allen, P. and McGlade, J., 1986. Dynamics of discovery and exploitation: The case of the Scotian shelf groundfish fisheries. *Canadian Journal of Fisheries and Aquatic Science*, 43, pp.1187–1200.

Allen, P. and McGlade, J., 1987. Modelling complex human systems: A fisheries example. *European Journal of Operational Research*, 30, pp.147–167.

Beddington, J. and May, R., 1977. Harvesting natural populations in a randomly fluctuating environment. *Science*, 197, pp.463–465.

Begon, M., Harper, J.L., and Townsend, C.R., 1996. *Ecology* (3rd edn). Oxford: Blackwell Science Ltd.

Berck, P. and Perloff, J., 1984. An open-access fishery with rational expectations. *Econometrica*, 52(2),489–506.

Bjorndal, T. and Conrad, J.M., 1987. The dynamics of an open access fishery. *Canadian Journal of Economics*, 87, 74–85.

Bulte, E. and Horan, R., 2001. Habitat conservation, wildlife extraction and agricultural expansion. Paper presented at EAERE 2001 conference University of Southampton, U.K., held 28–30 June 2001. (Discussion paper, submitted to JEEM).

Bulte, E.H., 2003. Open access harvesting of wildlife: The poaching pit and conservation of endangered species. *Agricultural Economics*, 28, pp.27–37.

Carr, M. and Reed, D., 1993. Conceptual issues relevant to marine harvest refuges: Examples from temperate reef fishes. *Canadian Journal of Fisheries and Aquatic Science*, 50, pp.2019–2028.

Clark, C., 1990. *Mathematical Bioeconomics: The Optimal Management of Renewable Resources* (2nd edn). New York: John Wiley and Sons.

Clark, C., 1973. Profit maximisation and the extinction of animal species. *Journal of Political Economy*, 81, pp.950–961.

Conrad, J., 2004. Renewable resource management. *Encyclopedia of Life Support Systems*. Paris: UNESCO.

Conrad, J., 1995. Bioeconomic models of the fishery. In D. Bromley (Ed.), *The Handbook of Environmental Economics*. Oxford: Blackwell, pp.405–432.

Crookes, D.J., 1997. *A Comparative Analysis and Critique of Selected Bioeconomic Models*. M.Sc. Dissertation, Department of Economics, University of Warwick, Coventry, U.K.

Crookes, D.J., 2018. Does the construction of a desalination plant necessarily imply that water tariffs will increase? A system dynamics analysis. *Water Resources and Economics*, 21, pp.29–39.

Damania, R., Stringer, R., Karanth, K.U., and Stith, B., 2003. The economics of protecting tiger populations: Linking household behavior to poaching and prey depletion. *Land Economics*, 79(2), pp.198–216.

Dendrinos, D.S. and Mullally, H. 1981. Evolutionary patterns of urban populations. *Geographical Analysis*, 13(4), pp.328–344.

Dendrinos, D.S. and Mullally, H., 1983. Optimum control in nonlinear ecological dynamics of metropolitan areas. *Environment and Planning, A*, 15, pp.543–550.

Fowler, C., 1981. Comparative population dynamics in large mammals. In C. Fowler and T. Smith (Eds), *Dynamics of Large Mammal Populations*. New York: Wiley and Sons.

Fowler, C., 1984. Density dependence in cetacean populations. *Report of the International Whaling Commission*, Special issue 6, pp.373–379.

Fisher, A.., Krutilla, J., and Cicchetti, C., 1972. The economics of environmental preservation: A theoretical and empirical analysis. *American Economic Review*, 62, pp.605–619.

Fox, W., 1970. An exponential surplus-yield model for optimising exploited fish populations. *Transactions of the American Fishery Society*, 99(1), pp.80–88.

Gordon, H., 1954. The economic theory of a common property resource: the fishery. *Journal of Political Economy*, 62, pp.124–142.

Harrison, S., 1994. Metapopulations and conservation. In P. Edwards, R. May, and N. Webb (Eds), *Large-Scale Ecology and Conservation Biology*. Cambridge, UK: Blackwell Scientific Publications.

Hofer, H., Campbell, K., East, M., and Huish, S., 2000. Modeling the spatial distribution of the economic costs and benefits of illegal game meat hunting in the Serengeti. *Natural Resource Modeling*, 13(1), pp.151–177.

Hortsthemke, W. and Lefever, R., (1984). *Noise-Induced Transitions: Theory and Applications in Physics, Chemistry and Biology*. Germany: Springer-Verlag.

Ibáñez, J., Valderrama, J.M., and Puigdefábregas, J., 2008. Assessing desertification risk using system stability condition analysis. *Ecological Modelling*, 213(2), pp.180–190.

Levins, R., 1970. Extinction. *Lectures on Mathematics in the Life Sciences*, 2, pp.77–107.

Liu, X. and Heino, M., 2014. Overlooked biological and economic implications of within-season fishery dynamics. *Canadian Journal of Fisheries and Aquatic Science*, 71, pp.181–188.

MacArthur, R. and Wilson, E., 1967. *The Theory of Biogeography*. Princeton, NJ: Princeton University Press.

Mackay, B. and Acheson, J., 1987. *The Question of the Commons: The Culture and Ecology of Communal Resources*. Tucson, AZ: University of Arizona Press.

May, R., 1994. The effects of spatial scale on ecological questions and answers. In P. Edwards, R. May, and N. Webb (Eds), *Large-scale Ecology and Conservation Biology: The 35th Symposium of the British Ecological Society with the Society for Conservation Biology*, University of Southampton, 1993, pp.1–9. Oxford: Blackwell Scientific Publications.

May, R., Beddington, J., Clark, C., Holt, S., and Laws, R., 1979. Management of multispecies fisheries. *Science*, 205(4403), pp.267–277.

Milner-Gulland, E. and Mace, R., 1998. *Conservation of Biological Resources*. Oxford: Blackwell Science.

Milner-Gulland, E., Shea, K., Possingham, H., Coulson, T., and Wilcox, C., 2001. Competing harvesting strategies in a simulated population under uncertainty. *Animal Conservation*, 4, pp.157–167.

Milner-Gulland, E. and Leader-Williams, N., 1992. A model for incentives for the illegal exploitation of black rhinos and elephants: Poaching pays in Luangwa Valley, Zambia. *Journal of Applied Ecology*, 29, pp.388–401.

Opsomer, J.-D. and Conrad, J., 1994. An open-access analysis of the northern anchovy fishery. *Journal of Environmental Economics and Management*, 27, pp.21–37.

Orishimo, I., 1987. An approach to urban dynamics. *Geographical Analysis*, 19(3), pp.200–210.

Pella, J. and Tomlinson, P., 1969. A generalised stock production model. *Bulletin of the Intertropical Tuna Commission*, 13, pp.421–458.

Puliafito, S.E., Puliafito, J.L., and Grand, M.C., 2008. Modeling population dynamics and economic growth as competing species: An application to CO_2 global emissions. *Ecological Economics*, 65(3), 602–615.

Sanchirico, J. and Wilen, J., 1999. Bioeconomics of spatial exploitation in a patchy environment. *Journal of Environmental Economics and Management*, 37, pp.129–150.

Sanchirico, J., and Wilen, J., 2000. *Dynamics of Spatial Exploitation: A Metapopulation Approach*. Discussion Paper 00–25-REV. Resources for the Future, Washington, D.C.

Schaefer, M., 1954. Some aspects of the dynamics of populations important to the management of the commercial marine fisheries. *Bulletin of the Inter-American Tropical Tuna Commission*, 1(2), pp.27–56.

Schaefer, M., 1957. Some considerations of population dynamics and economics in relation to the management of the commercial marine fisheries. *Journal of the Fisheries Research Board of Canada*, 14(5), pp.669–681.

Schultz, C. and Skonhoft, A., 1996. Wildlife management land-use and conflicts. *Environment and Development Economics*, 1, pp.265–280.

Skonhoft, A., 1998. Resource utilization, property rights and welfare. Wildlife and the local people. *Ecological Economics*, 26, pp.67–80.

Smith, V., 1968. Economics of production from natural resources. *American Economic Review*, 58, pp.409–431.

Smith, V., 1969. On models of commercial fishing. *Journal of Political Economy*, 77, pp.181–198.

Sugden, R., 1989. Spontaneous order. *Journal of Economic Perspectives*, 3(4), pp.85–97.

Sutton, W. and Jarvis, L., 1998. Optimal management of multi-value renewable resources: an application to the African Elephant. 1998 American Agricultural Economics Association meeting selected paper.

Swallow, S., 1990. Depletion of the environmental basis for renewable resources: the economics of interdependent renewable and non-renewable resources. *Journal of Environmental Economics and Management*, 19, pp.281–296.

Swanson, T., 1994. *The International Regulation of Extinction*. London: MacMillan Press.

Wilen, J., 1976. *Common Property Resources and the Dynamics of Overexploitation: The Case of the North Pacific Fur Seal*. Paper No. 3 in the Programme in Resource Economics, Department of Economics, University of British Columbia.

3 Co-evolutionary models and system dynamics modelling

3.1 Introduction

System dynamics modelling is an equation-based modelling platform where feedback most commonly characterises the interactions between the components. Therefore, the modelling framework lends itself very well to co-evolutionary models, for example through the use of the predator–prey suite of equations. Although these equations may be constructed in Excel and other modelling platforms, the system dynamics platform using software such as Vensim has the following advantages:

1. The interactions between the different components of the model are represented visually in the system dynamics platform using a stock–flow diagram, enabling transparency.
2. Various validation processes may be employed on a system dynamics model that cannot be conducted in a spreadsheet (such as unit consistency checks, structure verification tests, sensitivity analysis and extreme conditions tests). These tests increase the confidence in the robustness of the model. A further elaboration of these validation tests will be provided later in the chapter.
3. System dynamics models may be used to estimate the values of unknown parameters in the model through Monte Carlo simulation. Although Excel Solver can also fulfil this role, Crookes and Blignaut (2019) demonstrate that the Monte Carlo simulation techniques in Vensim provide an improved means of deriving these parameter estimates in these predator–prey models characterised by oscillating behaviour between predator and prey.
4. System dynamics models are easier to replicate than Excel models, since the equations are usually provided in the published work.

In spite of these advantages, co-evolutionary models using predator–prey interactions are not so common in the system dynamics community. The

DOI: 10.4324/9781003247982-3

purpose of this book is to highlight some of the ways in which this technique can be used, by highlighting the steps in the modelling process, the different equations from the population modelling literature that can be incorporated and the steps in the validation literature.

3.2 Steps in the modelling process

The following steps are employed in using co-evolutionary models in simulation [these steps are loosely based around Swartzman and Kaluzny (1987)]. It should be noted that, following Sterman (2000), model development is an iterative process and may involve revisiting earlier steps.

1. Identify and conceptualise a problem and define model objectives
2. Define the variables of interest
3. Identify the functional form that characterises the interactions between the variables
4. Model development. In some cases, parameters may be based on scant information and best guesses are provided to ensure that the model converges on reasonable dynamic outcomes
   - Construct a stock–flow diagram with the key elements in the model and their interactions
   - Add equations to each of the elements in the model
   - Add values for the constants in the model based on literature estimates of these values
   - When the value of a constant is not known, assign low values to unknown parameters so that they do not affect the dynamics of the model (these are later refined through optimisation)
   - Develop charts to show the key dynamics of the system over time
5. If historical data is available, use curve fitting to refine estimates of parameters (a further elaboration on conducting this in Vensim is given in Chapter 4)
   - Tools such as SyntheSim (Vensim) are used to get approximate values for unknown parameters by moving sliders of different unknown parameters until an approximate fit with the historical data is achieved
   - Use optimisation to refine values of unknown constants that achieve the best fit with the historical data
6. Conduct validation checks on the structure of the model. Key questions to ask are the following: Is the model structure used reputable? Do the parameter values estimated through optimisation align with common sense/the literature? Are the dimensions (units) of the model consistent?

- Conduct sensitivity analysis on structure of the model, as well as extreme value tests (for example by using Reality Check in Vensim)
- Further tests are used to validate the behaviour of the model and consider the policy implications

7. Model use. The model may then be used
   - To answer what-if type policy questions
   - To make forecasts of stock variables
   - To enhance knowledge on the values of parameters
   - To make inferences about the behaviour of entities in the model
   - As input into other models, such as game-theoretic problems

In the next section, we will consider the validation of these system dynamics models.

3.3 Validation

3.3.1 Introduction

Model validation is essential if a Lotka–Volterra (LV) model is to be used for forecasting (Ginzburg and Jensen 2004). It is the process of developing confidence in the model. For example, for econometric analysis numerous tests are used to validate a model (e.g. test of model and parameter significance, adjusted R-squared values). Most commonly, LV models are validated by comparing model results with actual data, for example through the use of an appropriate statistical measure (Armstrong 2001), a visual plot or by comparing with other models (Li et al. 2017). However, there are a number of other tools that may also be employed.

One useful way of validating a predator–prey model is through the tools offered by different software and modelling packages. For example, a stock–flow diagram may be constructed in system dynamics software such as Vensim in order to provide a visual representation of the system, which makes it possible to see how the different components in the model interact with one another. The structure of the model may be assessed by using regression analysis (Capello and Faggian 2002). Dimensional consistency is another tool offered by system dynamics software and is used to investigate whether the units are consistent. Sensitivity analysis is also employed in a variety of modelling packages in order to evaluate the predictions of the model, for example by using Monte Carlo simulation. Other methods (such as Bayesian techniques) are used to estimate the effect of extreme conditions on the model (Crookes 2017). Finally, optimisation is used to estimate unknown parameter values by comparing model simulations with actual data. Although these techniques require advanced software that are not accessible to every LV modeller, they nonetheless provide a rigorous means of validating these models.

Given the potential plethora of validation techniques that can be used, a review of the most commonly cited techniques by five prominent system dynamics modelling authors is provided in order to determine which validation techniques are most common. The results from this analysis (Table 3.1) indicate that structure verification, parameter verification, dimensional consistency, boundary adequacy, extreme conditions, surprise behaviour, sensitivity analysis and behaviour reproduction are *sine qua non*. This is not to discount the importance of other tests listed in the table, whose use would be dictated by the nature and complexity of the system dynamics model in question. However, these eight tests provide a minimum requirement for system dynamics modellers. Each of these tests are described in the next section.

3.3.2 *Descriptions of validation tests*

3.3.2.1 *Structure verification*

Structure verification involves comparing the model structure with structures prevalent in real-world situations, or patterns of relationships found in the literature, from established models or through expert opinion. In the case of a predator–prey model, this would mean some form of external validation that a predator–prey relationship does exist, either from the literature or expert opinion. Structure verification in this case represents the LV model that provides the best representation of reality.

3.3.2.2 *Parameter verification*

In the same way that model structure is compared with real-world situations, model parameters (constants) are also validated by comparing with actuality. Parameter verification is closely related to structure verification in that different parameter values influence the outcome of the model structure.

Parameter values are ideally obtained from published literature sources or are endogenously generated by the model (for example by optimisation). Personal interviews and meetings with a range of experts are also appropriate.

3.3.2.3 *Dimensional consistency*

The dimensional consistency test involves the analysis of a model's equations in order to test whether the model's dimensions (units) are consistent.

Table 3.1 Summary of validation tests conducted by different authors

Forrester and Senge (1980)	Richardson and Pugh (1981)	Sterman (2000)	Schwaninger and Groesser (2009)	Hill (2010a, 2010b)	Number used (out of 5)
Structure verification	Face validity	Structure assessment	Structure examination	Structure verification (implied)	5/5
Parameter verification	Parameter values	Parameter assessment	Parameter examination	Parameter verification (implied)	5/5
Dimensional consistency	Dimensional consistency	Dimensional consistency	Dimensional consistency	Dimensional consistency	5/5
Boundary adequacy	Boundary adequacy	Boundary adequacy	Boundary adequacy	Boundary adequacy	5/5
Extreme conditions	Extreme conditions	Extreme conditions	Extreme conditions		4/5
Surprise behaviour	Surprise behaviour	Surprise behaviour	Surprise behaviour		4/5
Behaviour sensitivity	Parameter sensitivity	Sensitivity analysis	Behaviour sensitivity		4/5
Behaviour reproduction		Behaviour reproduction	Behaviour reproduction		3/5
Behaviour anomaly		Behaviour anomaly	Behaviour anomaly		3/5
Family member		Family member	Family member		3/5
		Integration error	Integration error	Integration error	3/5
Behaviour prediction			Behaviour anticipation		2/5
		System improvement	System improvement		2/5
Extreme policy					1/5
				Mass balance check	1/5
			Loop dominance		1/5
			Turing test		1/5

Notes: Tests listed are those discussed in the source. Hill (2010b) may not represent an exhaustive list of validation methods employed either by himself or Ventana Systems UK. Schwaninger and Groesser (2009) also discuss a number of context-related tests that are not discussed by any of the other authors.

Coyle and Exelby (2000:35) also regard the dimensional consistency test as being *sine qua non*.

Most good system dynamics modelling software packages provide a means of testing the dimensional consistency of the model. Dimensional consistency is checked throughout the building of the model, and the final model units should satisfy the requirements.

3.3.2.4 Boundary adequacy

Boundary adequacy comprises three separate tests, namely the structure boundary adequacy test, the behaviour boundary adequacy test and the policy boundary adequacy test, that contain essentially the same logic (Sterman 2000). The structure boundary adequacy test considers whether all the important elements of a structure are contained in the model and what level of aggregation is appropriate. The behaviour boundary adequacy test asks whether model behaviour would change significantly if boundary assumptions were changed. The policy boundary adequacy test investigates whether policy recommendations would change as a result of a change in the model boundary.

For the structure boundary adequacy test, exogenous variables are examined in order to ascertain whether they are complete and also whether there are certain exogenous variables that should be endogenous. A number of techniques may be used to accomplish this. Using models from the literature ensures that appropriate structures are followed. Expert opinion may also be used in order to determine whether important structural characteristics were omitted from the model, and the behaviour responses are tested by investigating what changes in behaviour would result as a consequence of a change in model structure (structure boundary adequacy). The effect of policy recommendations is also tested in this way (policy boundary adequacy). It is important to assess whether changes in the model structure, behaviour and policy are consistent with real-world considerations.

3.3.2.5 Extreme conditions

This test assigns extreme but realistic values to parameters and investigates whether the model responds in an expected manner. Various extreme conditions are applied to parameters in order to ascertain how the system responds.

3.3.2.6 Surprise behaviour

A surprise behaviour is a model behaviour that is not anticipated by the model analysts. Sometimes this is due to a formulation flaw in the model, and other times this may lead to an identification of behaviour previously unrecognised in the real-world system. In the latter case, confidence in the model's usefulness is strongly enhanced. If a test for surprise behaviour leads to unexpected results, the modeller must understand the causes of the unexpected behaviour in the model.

3.3.2.7 Sensitivity analysis

Sensitivity analysis tests the robustness of the model to underlying assumptions. Following Sterman (2000), three types of sensitivity are distinguished: numerical sensitivity, behavioural sensitivity and policy sensitivity. Numerical sensitivity is a feature of the model, while behavioural sensitivity and especially policy sensitivity are important and require testing. Sensitivity analysis involves testing the sensitivity of model results to changes in parameter values.

3.3.2.8 Behaviour reproduction

Behaviour reproduction is the only test that is included in the discussion that scored a three in our review of tests (see Table 3.1), but is nonetheless useful to include in system dynamics models. This test involves utilising qualitative and quantitative measures for comparing how best the model is able to replicate the actual behaviour of the system. Quantitative methods include statistical measures such as using the coefficient of determination (R^2), mean absolute error (MAE), mean squared error (MSE), mean absolute percentage error (MAPE) and Theil's inequality statistic to investigate how much the simulation model deviates from actual values.

3.4 System dynamics as a tool for machine learning

System dynamics may also be used for machine learning, particularly in the context of forecasting. It involves splitting the data into a training component and a testing component. The training component of the model is used to calibrate the model with the historical data, and the forecast accuracy of the model is then assessed by comparing it with the testing component of the actual data set using an appropriate statistic. This approach is therefore similar to the behavioural reproduction validation test proposed earlier, but in this case involves an additional step. System dynamics can be used in conjunction with machine learning techniques, such as optimisation, regression analysis, and regularisation as well as Bayesian methods, in order to make predictions.

In the next chapter, we discuss how to use Monte Carlo methods in Vensim to estimate the values of unknown parameters.

References

Armstrong, J.S., 2001. Evaluating forecasting methods. In J. Scott Armstrong (Ed.), *Principles of Forecasting: A Handbook for Researchers and Practitioners.* Boston, MA: Springer.

Ginzburg, L. R. and Jensen, C. X. J., 2004. Rules of thumb for judging ecological theories. *Trends in Ecology & Evolution*, 19, pp.121–126.

Capello, R. and Faggian, A., 2002. An economic-ecological model of urban growth and urban externalities: Empirical evidence from Italy. *Ecological Economics*, 40, pp.181–198.

Coyle, G. and Exelby, D., 2000. The validation of commercial system dynamics models. *System Dynamics Review*, 16(1), pp.27–41.

Crookes, D.J., 2017. Does a reduction in the price of rhino horn prevent poaching? *Journal for Nature Conservation*, 39, pp.73–82.

Crookes, D.J. and Blignaut, J.N., 2019. An approach to determine the extinction risk of exploited populations. *Journal for Nature Conservation*, 52, p.125750.

Forrester, J.W. and Senge, P.M., 1980. Tests for building confidence in systems dynamics models. In Legasto, A.A., Forrester, J.W. and Lyneis, J.M. (Eds), *System Dynamics. TIMS Studies in the Management Sciences* 14. Amsterdam: North-Holland Publishing Company, pp.209–228.

Hill, A., 2010a. *System Dynamics Using Vensim*. Salisbury: Ventana Systems UK Ltd.

Hill, A., 2010b. Personal interview. 17 September, Salisbury, United Kingdom.

Li, S., Chen, H., and Guofang Zhang, G., 2017. Comparison of the short-term forecasting accuracy on battery electric vehicle between modified bass and Lotka-Volterra model: A case study of China. *Journal of Advanced Transportation*, 2017, Article ID 7801837.

Richardson, G.P. and Pugh, A.L., 1981. *Introduction to System Dynamics Modelling with DYNAMO*. Portland, OR: Productivity Press.

Schwaninger, M. and Groesser, S., 2009. System dynamics modelling: validation for quality assurance. In Meyers, R.A. (Ed.), *Encyclopedia of Complexity and Systems Science. Volume 9*. New York: Springer.

Schwartzman, G.L. and Kaluzny, S.P., 1987. *Ecological Simulation Primer*. New York: Macmillan.

Sterman, J.D., 2000. *Business Dynamics. Systems Thinking and Modelling for a Complex World*. Boston, MA: Irwin McGraw-Hill.

4 Using numerical methods to estimate unknown parameters in co-evolutionary models

4.1 Basic model

Following are the steps in estimating unknown biological parameters in the predator–prey model, with historical data known. [Note: This model is built in Vensim and calibrated using Vensim's optimisation routine. This model cannot be built using Vensim PLE (the free version of Vensim), which does not include optimisation.] This model was built in Vensim DSS Version 6.4b and calibrated using the Markov chain Monte Carlo (MCMC) simulation method.

Data for the model is from the Hudson Bay Company (1900–1920) and was printed in Purves et al. (1992) and reproduced on the website of San Diego State University and Joseph M. Mahaffy (2010). The reader is directed to this website in order to obtain the data, or contact the present author for further information. The data are for furs collected, but it is generally assumed that these furs are representative of the underlying populations of lynx and hares over that period (SDSD and Mahaffy 2010).

1. Open an Excel spreadsheet, save as 'data' (also rename the current tab as data). Save as old version of Excel (1997–2003) (may not be necessary for newer versions of the software).
2. The first column must be the time variable; other columns of data must correspond with stocks in the model.
3. Save data as 'data.xls' (1997–2003 version).
4. We are going to use the basic Lotka–Volterra predator–prey equations and then use Vensim to solve for the unknown parameter values by calibrating the model with the historical data:

$$H(x, y) = rx - \alpha xy$$

DOI: 10.4324/9781003247982-4

$$L(x, y) = \beta xy - my$$

where $H(x,y)$ is the growth rate of the hares (prey), r is the growth rate and α is the mortality coefficient. $L(x,y)$ is the growth rate of the predator (lynx), $-m$ is the mortality of the predator, and β is the growth of the predator, dependent on the abundance of the prey x. We will now create this model in Vensim. This simple model assumes no density dependence.

5. Create a new Vensim model (call it lynx).
6. Open Vensim.
7. Click file ---. New model. A window should open.
8. Enter initial time = 1900.
9. Enter final time = 1920.
10. Enter units for time select 'year'.
11. Click OK.
12. Add a stock variable (box) for hares (x).
13. Add a rate variable going in and call it growth hares.
14. Add a rate variable coming out and call it mortality hares.
15. Your diagram should look like that in Figure 4.1.
16. Add a variable below growth hares and call it r.
17. Add an arrow from r to growth hares.
18. Add a variable below mortality hares and call it alpha.
19. Add an arrow to connect alpha to mortality hares.
20. Your model should now look like Figure 4.2.
21. Now add an arrow from hares to growth hares and one from hares to mortality hares (click on hares, then click on a space between the two variables and then click on growth hares. That should make a nice arced line). Your model should now look like Figure 4.3.
22. Create a similar diagram for the lynx (the easiest is to copy and paste the above by highlighting using the hand tool and then clicking copy (ctrl c) and clicking paste (ctrl v)) (select the replicate option). You will then need to drag the copied portion down to below the hares sub-model using your mouse. Then rename the stock (box) variable lynx, growth hares rename as growth lynx, mortality hares rename as mortality lynx, and

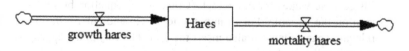

Figure 4.1 Stock-flow diagram for hares.

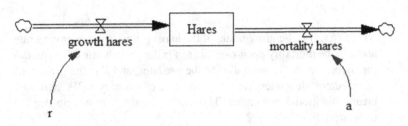

Figure 4.2 Stock-flow diagram for hares with growth coefficient *r* and mortality coefficient alpha (*a*) added.

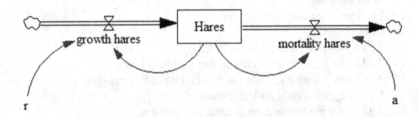

Figure 4.3 Stock-flow diagram for hares with feedback with growth and mortality added.

rename r as beta and alpha as m (click on the variable button and then on the variable to rename it). Your diagram should now look Figure 4.4.

23. Finally, draw an arrow from hares to growth lynx, and from lynx to mortality hares. This is because an increase in the population of hares increases the growth of lynx, and an increase in the population of lynx increases the mortality of hares, which is what the data show. Your figure should now look like Figure 4.5.

24. Click file save as and save in your model folder as 'lynx'.

25. Now we are going to add equations and values for the variables in the model.

26. We utilise the initial parameters in the model, m = 0.08, alpha = 0.001, r = 0.02, beta = 0.00002 (as some introductory teaching texts propose).

27. Insert these values for the model. [Click on the equation button, and then on the variable. For example, click on m and enter 0.08 in the relevant space. One can also insert the units (1/year) in the appropriate unit cell.] Click OK.

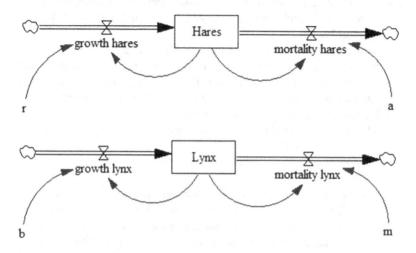

Figure 4.4 Stock-flow diagram for hares and lynx with feedback and growth and mortality coefficient added.

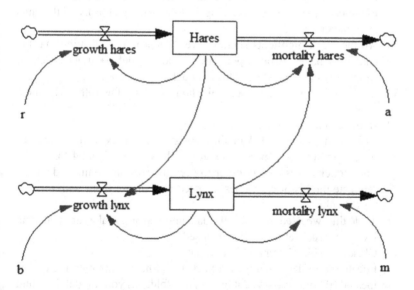

Figure 4.5 Stock-flow diagram for hares and lynx with predator between lynx and hares added.

28. Do the same for alpha (value = 0.001, units = 1/(number*year)), r (value = 0.02 and units 1/year), and beta (value = 0.00002 units 1/(number*year))

29. Now we must add the initial values of hares and lynx, which we get from our data. The initial value in 1900 was 30 (×1000) for hares and 4 (×1000) for lynx. We will ignore the thousands for now. Click on the hares stock. A dialogue box should pop up. Insert 30 in the initial value location and under units type in number.

30. Click OK and then do the same for lynx (initial value = 4 and units = number). Click OK.

31. We now have to modify the rates equations.

32. Click growth hares.

33. In the dialogue box, you will see variables hares and r on the far right-hand side. These are input variables. You will multiply them together, as per the Lotka–Volterra formula. Double-click on r and it should appear in the equation editor. Then type *, then click on hares. Finally add number/year in the units.

34. Click OK to accept.

35. Do the same for all the other rates variables (mortality hares, growth lynx, mortality lynx). In all cases there should be two or three variables to multiply together (in the far right of the dialogue box), and the units are number/year.

36. All the black highlights should have disappeared, indicating that the equations have all been populated. Click model → units check. It should say units ok. If not, check model entries again.

37. Click model → check model. This should also say OK (otherwise there is an error).

38. If model and units OK, click simulate.

39. Hold down shift and click on the hares stock and lynx stock. Then click on the graph button on the left panel of the screen (Figure 4.6).

40. The model works, but in order to see how well the simulated values replicate the historical values we need to compare them with the historical values.

41. To do this, we need to import the data into Vensim. Make sure your data Excel spreadsheet is saved and closed.

42. Click model in Vensim → import dataset.

43. Locate the folder where your Excel spreadsheet containing the data is located (if you have saved it in the same folder as your model, it should be visible).

44. Click on the data spreadsheet and click open. A dialogue window should open.

45. Click on down for the time data, as the data are in columns, not rows.

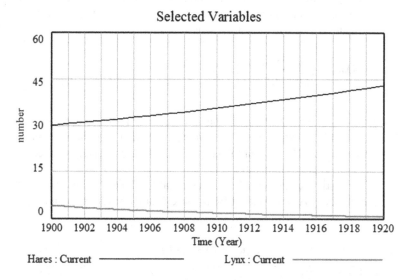

Figure 4.6 Graphical plot of the co-evolution of lynx and hares over time.

46. Click OK.
47. Now the data have been imported into Vensim. Click OK.
48. Take note of the file name that it is imported into. If you remembered to change the data.xls tab to data, then your data is imported as data .vdf. Otherwise it will probably be sheet1.vdf.
49. If you click simulate and then shift, and select hares and lynx, and click the graph button, you should see the simulated data and the historical data. It should look something like Figure 4.7.
50. The historical data and the model simulated populations diverge quite noticeably.
51. We can use the formula for mean annual percentage error (MAPE) to estimate how well the model fits with the data.
52. Click on model → export dataset.
53. Select Current.vdf as the file and click open. Under "export to" type Baseline. Then select Excel file as the destination. The select time running down.
54. Click OK.

55. These give the values for your simulation runs, along with the parameter values. One can then copy the historical data into the folder and estimate MAPE using the following formula:

$$MAPE = \frac{1}{n}\sum_{t=1}^{n}\left|\frac{\hat{x}_t - x_t}{\hat{x}_t}\right| \times 100$$

where $\hat{x}_t$ is the actual population abundance of lynx and hares in time t, x_t is the simulated population abundance, and n is the number of observations. The criteria for assessing the forecast accuracy based on the MAPE is given in Table 4.1.

56. If we estimate the MAPE we get a value of 86.3% for hares and 99.4% for lynx. It is evident from Table 4.1 that both of these are poor

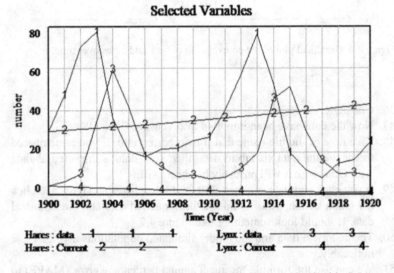

Figure 4.7 Graphical plot of the lynx and hares data compared with the current model simulated results (baseline model).

Table 4.1 Forecast accuracy of the model based on MAPE

MAPE	Forecast accuracy
0–10%	Highly accurate
10–20%	Good
20–50%	Reasonable
50–100%	Poor

forecasts. The underlying parameter values that we assumed are evidently not very realistic.

57. Save this Excel spreadsheet using the latest version of Vensim.

4.2 Calibrated model

58. We can use calibration to attempt to get a better fit with the historical data by varying the estimates of the parameters alpha, beta, r and m. Ultimately, we are after more realistic estimates of the biological parameters alpha, beta, r and m. This is what we are going to do now.
59. Click the optimise button.
60. Enter lynx as the payoff definition and click ok.
61. Click add and then select payoff type calibration and select hares and keep the weight as 1. Click OK.
62. Click add again and select calibration, and then select lynx and also keep the weight as 1. Click OK.
63. Click next.
64. From the dropdown menu, change optimiser from Powell to MCMC.
65. Click add constant.
66. Select alpha and click OK.
67. Enter 0 as the minimum value and 0.01 as the maximum value.
68. Click add constant again.
69. Select beta from the list and click OK.
70. In the minimum field, enter 0 and at the maximum enter 0.05.
71. Click add constant and select m and click OK.
72. Select the minimum as 0 and the maximum as 1.7.
73. Finally click add constant and select r and click OK.
74. Set the minimum as 0 and the maximum as 0.3.
75. Click next.
76. Click on the Data sources button and select the data.vdf file you created (or whatever your file with the data is called).
77. Click finish.
78. Let the model run for a minute or so and then click stop.
79. It will ask you if you want to do a final simulation. Click yes.
80. Hold down shift and click on the hares stock and the lynx stock variable. Both should be highlighted. Then click on the graph button on the left of the screen. You should get Figure 4.8.
81. You will notice that the model gives a better fit with the historical data compared with the baseline model. To see how much better, we can calculate the MAPE for the data.
82. Click model → export dataset and click Current.vdf and open.

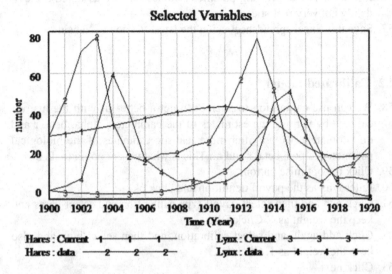

Figure 4.8 Graphical plot of the lynx and hares data compared with the current model simulated results (lynx model).

83. Under export to type in lynx and select Excel and time running down as before.
84. Click OK. You can now copy the historical data into this new spreadsheet and calculate MAPE once again.
85. You should now get an MAPE for hares of 56.53% and for lynx of 85.92%. According to the accuracy table, the simulated values for hares and the lynx, although still a poor forecast of the historical data, is nonetheless an improvement over the baseline.
86. It is not perfect, but this can be improved through revising the underlying functional form (for example by including density dependence).
87. Save this file to the latest version of Excel (or your data will be lost).
88. Hold down shift and select the constants r, a, b and m in the stock-flow diagram (these can also be viewed in your lynx spreadsheet). They should be highlighted. Then click the table button in the left-hand panel.
89. These are parameters that gave the best fit in the MCMC optimisation routine, based on the constraints given to the optimisation parameters. These can then be inserted into the model using the equation editor, thereby replacing the old values. The model can then be used to forecast future values of the population or used to estimate steady-state values of the populations in question. Inference can also be made on the "best" model that fits the data, in order to better understand the nature of the

Table 4.2 Comparison between old and new values

Parameter	Old value (baseline model)	New value (lynx model)
Alpha	0.001	0.00559
Beta	0.00002	0.04936
m	0.08	1.679
r	0.02	0.05798
MAPE	86.3% for hares and 99.4% for lynx	56.53% for hares and 85.92% for lynx

biological interactions between the predator and the prey. Table 4.2 highlights the differences between the "old" and "new" values.

90. We observe that all the parameters are higher than under the baseline.
91. Future work can attempt further iterations in order to attempt to improve the MAPE of the model further, thereby getting a better understanding of the nature of the interactions between the species in the model and their environment.

4.3 Exercises

1. Using the equations

$$H(x,y) = rx\left(1 - \frac{x}{K}\right) - \alpha xy$$

$$L(x,y) = sy\left(1 - \frac{y}{L}\right) + \beta xy - ms$$

repeat the preceding steps (saving the model as Lynx 1). Do you get an improved MAPE for hares and lynx population dynamics?

2. Now let's try

$$H(x,y) = rx\left(1 - \left(\frac{x}{K}\right)^{z1}\right) - \alpha xy$$

$$L(x,y) = sy\left(1 - \left(\frac{y}{L}\right)^{z2}\right) + \beta xy - ms$$

Save the model as Lynx 2. Does this result in an improvement in MAPE over question 1?

References

de Andrade, P.R. and Câmara, G., 2009. *Predator Prey Models*. Brazil: Instituto Nacionale De Pesquisas Espaciais.

Purves, W.K., Orians, G.H., and Heller, H.C., 1992. *Life: The Science of Biology*. Salt Lake City, UT: W.H. Freeman.

SDSU and Mahaffy, J.M., 2010. *Math 636 - Mathematical Modeling, Fall Semester, 2010, Lotka-Volterra Models*. San Diego State University. Available at: https://jmahaffy.sdsu.edu/courses/f09/math636/lectures/lotka/qualde2.html#Lotka-Volterra (accessed 04 March 2021).

5 Co-evolutionary models and rhino management

5.1 Introduction

Forecasting the future state of an entity is an important part of decision-making. For example, business managers need to forecast demand for their products, so that they can make informed decisions about production. Economists forecast economic variables such as inflation and growth in order to determine what macroeconomic policies are applicable. Environmental managers wish to know the future state of a biological population in order to determine the extinction risk of a species. In all of these cases, forecasting can help inform decision-makers on how best to manage the resources available to them.

Sherden (1998) estimated that the forecasting industry, broadly defined, was worth $200 billion, but these figures are dated. The weather services forecasting market (in other words, the sector that sells weather forecasting services) is expected to grow from $1.5 billion in 2020 to $2.3 billion in 2025, an annual growth rate of 9.3% (MarketsandMarkets 2020). The fact that forecasting is a growth industry is highlighted by the consumer trends forecasting industry, which has grown from almost nothing to be worth GBP36 million in 2011 (*The Telegraph* 2011), but this excludes the value of stock and market forecasts, which are also likely to be sizeable. Based on these data, we can reasonably assume that the forecasting industry is worth at least $1550 billion in today's (2021) prices. The forecasting industry is indeed big business.

Why is this important? Because forecasting is crucial to the survival of many entities. It is crucial to the survival of businesses. For example, weather forecasting is vital for the agriculture sector, the fishing industry, the energy sector and other industries dependent on the weather. Forecasting is crucial for consumers. For example, forecasting consumer trends helps companies to deliver the right products at the right time (WGSN 2021). Forecasting is also crucial for the financial sector. For example, market

DOI: 10.4324/9781003247982-5

forecasts are essential for making investment decisions, which could affect private individuals as well as pension funds, corporate investors and governments. It is safe to assume that all sectors of the economy are reliant on some form of forecasting.

But not only is forecasting critical to the survival of markets and the economy, but the survival of species is also critically dependent on accurate forecasts of population trends. For example, the International Union for Conservation of Nature (IUCN) Red List uses historical population trends as one of the assessments of threat to a biological population. But, historical trends are not necessarily indicators of future biological populations. There is a need for models that provide accurate long-term forecasts.

Lotka–Volterra (LV) models are utilised for forecasting a wide range of business, financial and microeconomic phenomena, including stock markets (Lee et al. 2005), competition between firms (Marasco et al. 2016), sales (Hung et al. 2017) and revenue growth in the retail sector (Hung et al. 2014). The models have immense potential for modelling intersectoral dynamics where the resource from one sector is used as an input in another sector. For example, Crookes and Blignaut (2016) model the dynamics of vehicle manufacturing using steel as an input (the prey). In many cases, the LV model performs the same or better than other comparable models (such as the bass model and neural networks; see e.g. Hung et al. 2014, and Crookes and Blignaut 2016). These models are also used to forecast macroeconomic phenomena. For example, Wu and Liu (2013) developed an LV model to forecast gross domestic product (GDP) and foreign direct investment (FDI).

At the same time, although these models have been used to model aquaculture (Cacho 1997; Ponce-Marbán et al. 2006), land-use change (Castro et al. 2018; Paul et al. 2019), the management of weeds (Jones et al. 2006; Grimsrud et al. 2008; McDermott et al. 2013) and many other applications, these types of models are underutilised tools for forecasting biological populations that are subject to exploitation, which is surprising given that these models largely emerged from the fisheries literature. A reason for these methods being less frequently used for forecasting bioeconomic phenomena is that the biological growth parameters (intrinsic growth rate, carrying capacity of the population or maximum population size) that are required for these models are frequently unknown in natural systems. The system dynamics modelling tool provides a means by which this limitation may be overcome. The unknown biological parameters may be estimated from trend data in biological populations and effort data (if available). System dynamics modelling software also provides a means for estimating these unknown parameters and validating these models. These two techniques [namely LV models coupled with the system dynamics (SD)

modelling platform] provide an improved means of forecasting bioeconomic phenomena.

This chapter provides an application of the coupled LV/SD technique for
forecasting bioeconomic phenomena. Rhino management is an important
case study. Rhino poaching in South Africa has escalated enormously in the
last ten years, leading to concerns over the possible survival of rhino populations. A predator–prey simulation model was developed in 2015 (Crookes
2017) based on data collected up until 2012. Using estimates of population abundance subsequent to 2012 (up until 2019), it is possible to test the
forecast accuracy of the model, particularly as it relates to rhino abundance,
over the ensuing seven years.

The chapter is laid out as follows. First, is a comparison of the forecast
of the Schaefer model with two other harvest functions, namely the Cobb–
Douglas harvest function and the Baranov harvest function. These LV models are then compared with a least-squares specification of the Schaefer
logistic model. After that, the best model is then selected to forecast rhino
abundance from 2012 to 2020 and compared with historical data on population and poaching data over that period.

5.2 The model

The equations of this LV model are based on Crookes's (2017) model of
rhino population dynamics. The form of LV model is as follows:

$$\frac{dx}{dt} = F(x) - mh \tag{5.1}$$

$$\frac{dE}{dt} = n(ph - cE) \tag{5.2}$$

where x is the prey species (in this case the rhino population) and E is the
poaching effort (the predator). $F(x)$ is the rhino growth function, p is the
price of rhino horn and c is the cost per unit capital. h is the harvest function,
assumed to follow the Cobb–Douglas production relationship:

$$h_t = aqE_t^\alpha x_t^\beta \tag{5.3}$$

where α and β are elasticities of substitution; q is the catchability coefficient,
which relates effort and stocks to harvests; and a is a scaling parameter. If a
$= \alpha = \beta = 1$, then the well-known Schaefer production function is obtained.
In this example, if rhino populations are abundant, profits are positive, and
since open access prevails, poachers (E) enter the game reserve as long as
ph exceeds cE. But as this occurs rhino populations decline, and h therefore

declines, so that poachers exits the game reserve. With poachers exiting the game reserve, rhino populations recover, resulting in a dynamic system.

The basic model (Equations 5.1 and 5.2) also includes a number of other parameters, such as the probability a poacher is detected and the magnitude of the penalty (Equation 5.4). $F(x)$ follows the Pella and Tomlinson (1969) specification (see Equation 5.5) with a density-dependent term, and the Schaefer production function is assumed. In the present study we extend this analysis and model three harvest specifications in total: (1) the (original) Schaefer (S) function; (2) a Cobb–Douglas (CD) function (Equation 5.3) and (3) a Baranov (BV) function (Equation 2.10) discussed earlier.

Poaching effort (E_t) evolves according to

$$E_{t+1} = E_t + n'\left(h_t^* - \frac{c}{p}E_t - bE_t \frac{f+p}{p} \right) \tag{5.4}$$

where n' is an adjustment coefficient; c and p are the cost of poaching and value of rhino horn sold, respectively; f is the fine; and b the probability of detection and conviction.

Rhino populations (x_t) evolve according to

$$x_{t+1} = x_t + rx_t - \frac{rx_t^{z+1}}{k^z} - mh_t^* \tag{5.5}$$

where r is the intrinsic growth rate, z is a Fowler density-dependent term, k is the carrying capacity and m is the mortality coefficient. The values of the parameters are given in Crookes (2017). The harvest coefficient h_t^* varies depending on whether the Schaefer model, the Cobb–Douglas or the Baranov catch equation is used. More particulars are given in the next section when the estimation methodology is discussed.

5.3 Least-squares estimation of production function

The first step is to estimate the value for a, α and β using the Cobb–Douglas production function (Equation 5.3). Historical estimates for rhino abundance, effort and harvests from 1990 to 2013 were employed for this purpose.

The ordinary least-squares (OLS) estimation gave the results shown in Table 5.1.

Given that aq and β are not significantly different from zero, we conclude that the least-squares estimation of the production function results in a trivial solution such that a harvest function is not needed for the model. We model this for our forecast model (Cobb–Douglas, $a = 0$), however,

Table 5.1 Regression results for Cobb–Douglas specification

	Coefficient	T stat	Statistic/data	Sig
aq	0.9944	−1.47984		n.s.
α	0.002169	5.497837		***
β	−0.00016	−0.10389		n.s.
Model F			34.49	***
Adj R^2			0.744	
n			24	

Notes: *** Significant at least at the 1% level; n.s. = not significant.

we also model an alternative specification where $a = 1$ in order to test the importance of the harvest function. Based on the results of the OLS estimation, the Cobb–Douglas harvest function is either

$$h_t = qE_t^\alpha \quad \text{or} \quad h_t = 0 \tag{5.6}$$

For the Baranov model (see Equation 5.7), the aforementioned Cobb–Douglas production function is also utilised for F_t, following Liu and Heino (2014). The rest of the parameter values for the model are reported in Crookes (2017). The models were constructed in Vensim, which allows for simultaneous feedback between the parameters in the model.

5.4 Historical data replication

The model simulations from the LV models are evaluated in two ways. Firstly, they are compared with a traditional econometric model of the Schaefer logistic model, and secondly they are compared with criteria proposed by Li et al. (2017). These evaluation steps are given in more detail here.

5.4.1 Cobb–Douglas catch equation

The Cobb–Douglas catch equation used in the model is given in Equation 5.6.

5.4.2 Schaefer production function

The Schaefer production function is a special case of the Cobb–Douglas catch equation, where $\alpha = \beta = 1$.

5.4.3 Baranov catch equation

Although the Cobb–Douglas production function is the most common in the open-access literature, there are other important harvesting functions in the fisheries literature. One example is the Baranov catch equation. The harvest at the end of the season is

$$h^b = \frac{F}{F+M}\left(1 - e^{-(F+M)T}\right)\bar{x}_0 \qquad (5.7)$$

where $\bar{x}_0$ is the initial population abundance; F and M are the fishing and natural mortality rates, respectively; and T is usually 1 (Liu and Heino 2014). Liu and Heino (2014) assume a Cobb–Douglas production function for F of the form

$$F_t = ah_t = aqE_t x_t^\beta \qquad (5.8)$$

where a is a scaling parameter and β captures the potential of a non-linear response of harvests to changes in abundance. Catch h^b therefore becomes

$$h_t^b = \frac{F_t}{F_t+M}\left(1 - e^{-(F_t+M)}\right)\bar{x}_0 \qquad (5.9)$$

5.4.4 Schaefer logistic model

Following Pella and Tomlinson (1969), the logistic growth model may be written in the form

$$z_{t+1} = (1+r)z_t - \frac{r}{qk}z_t^2 - qh_t + u_t \qquad (5.10)$$

where u_t is an error term and z_t is the catch per unit effort (CPUE). Under certain conditions this function may be estimated using OLS (Zhang and Smith 2011). The model assumes a Schaefer production function. More advanced estimation methods have been proposed. For example, Zhang and Smith (2011) propose a "CPUE like" estimator using a Cobb–Douglas production function, which is then estimated using maximum likelihood methods. In this case, however, we are interested in the output of the Schaefer model in order to compare it with our original model.

5.5 Adaptive expectations

The preceding discussion assumes that, as far as stocks are concerned, poachers are myopic, in other words that they only utilise current period information about abundance in order to determine their harvesting behaviour. Harvests under the adaptive expectations hypothesis are formulated based on expectations of future stock size:

$$H_t = A + \beta \hat{X}_{t+1} + u_t \tag{5.11}$$

Stocks adjust based on the following:

$$\hat{X}_{t+1} - \delta X_t = X_{t-1} - \delta X_{t-1} \tag{5.12}$$

where δ is an adjustment coefficient relating actual change to desired change. Substituting $\hat{X}_{t+1}$ into H_t and rearranging gives an equation that is estimable, however, estimating such an equation is problematic due to the presence of contemporaneous correlation between the lagged endogenous variable and the error term. A solution is to use the method of instrumental variables (IV) regression, however, a suitable instrument is needed. One method proposed by Liviatan involves using lagged values of an exogenous variable. From these values it is possible to compute long-term values for α and β (see Gujarati 2003), and it is then possible to determine the value of scale variable a.

5.6 Implications for poaching behaviour

Two broad categories of harvesting behaviour are considered (Table 5.2): firstly, harvesting decisions that are myopic versus those based on expectations of future prey abundance, and secondly, harvesting decisions that are based only on prey abundance versus those based on prey abundance and changes in poaching profitability. The myopic model simulates harvesting decisions based on constant effort focussed on prey abundance (model 1) as well as a variable effort based on profitability (model 2), while the adaptive expectations model (model 3) simulates the effect of harvesting decisions based on a variable effort and poaching decisions based on changes in profitability.

Table 5.2 Harvesting decisions modelled

		Constant effort	Variable effort
	Harvesting decisions ...	... based on changes in stocks	... based on changes in stocks and profitability
Myopic	... based only on current stocks	Model 1	Model 2
Adaptive expectations	... based on expectations of future stocks	—	Model 3

5.7 Mean absolute percentage error

The second stage of assessing the replication of the model with the historical data is conducted by comparing the LV and econometric model with data from 2010 to 2015 by calculating the mean absolute percentage error (MAPE). Although there are problems with using this measure to assess forecast accuracy (Tayman and Swanson 1999), it is still the most commonly used measure (Mentzer and Kahn 1995; Armstrong 2001). This measure is calculated as follows:

$$\text{MAPE} = \frac{1}{n}\sum_{t=1}^{n}\left|\frac{\hat{x}_t - x_t}{\hat{x}_t}\right| \times 100 \tag{5.13}$$

where $\hat{x}_t$ is the actual data on rhino abundance in time t, x_t is the simulated rhino abundance and n is the number of observations. The criteria for assessing the forecast accuracy based on the MAPE are given in Table 5.3.

5.8 Forecast accuracy

The forecast accuracy of the model is then assessed through a visual plot of the "best" model with the historical data, as well as comparing with other estimates in the literature.

In the next section, the results of the replication of the historical data and the forecast accuracy of the models are discussed, with reference to the econometric model, a visual plot and the MAPE measure.

5.9 Validation

Chapter 3 highlights the different ways in which an LV model may be validated. Here we validate the model by comparing it with the historical data. Three methods are employed. Firstly, we compare the model with actual data between 2010 and 2015 using the MAPE. Secondly, we compare the

Table 5.3 Forecast accuracy of the model based on MAPE

MAPE	Forecast accuracy
0–10%	Highly accurate
10–20%	Good
20–50%	Reasonable
50–100%	Poor

Source: Li et al. (2017).

estimates with other models. Thirdly, we utilise a visual plot of the data to compare actual versus forecasted values. Error bars are calculated based on uncertainty over the area under rhino management. In this way, a coefficient of variation (CV) of 30% was calculated. In the absence of available data to estimate the CV for numbers killed, the same CV is assumed. The smaller the CV, the greater weighting that data point and/or time series gets due to the higher precision associated with the data point. The CV for rhinos is relatively high, and indicates the uncertainty in the underlying spatial data and kill data rather than the population survey data, which is likely to be fairly accurate. In the next section, we use actual data to assess the forecast accuracy of the predator–prey model based on three different production functions.

5.10 Results

5.10.1 Parameter values

Most of the parameter values in the model are given in Crookes (2017) and are therefore not repeated here. The values obtained for a and β for the different models, which are summarised in Table 5.4, indicate that the Schaefer and Baranov models produce similar estimates for β. The value of a is much lower under the adaptive expectations model, which one expects as harvests have time to adjust to changes in stocks. But the most significant result from the analysis is that, over time when expectations fully adjust, the value of β is 1, indicating that the model reverts to the standard Schaefer and Baranov production functions.

5.10.2 Model 1: Myopic – current stocks only

The stock–flow diagram of model 1 (Figure 5.1) indicates that the myopic model is linear in the sense that there is no feedback between the different components of the model.

Table 5.4 Values of parameters under different harvesting functions

	a	*β*
Baranov production function		
Model 1: Myopic (constant *E*)	16.1	2.96
Model 2: Myopic (variable *E*)	15.55	2.96
Model 3: Adaptive expectations	1.33	1.00
Schaefer production function		
Model 1: Myopic (constant *E*)	13.33	2.96
Model 2: Myopic (variable *E*)	10.73	2.96
Model 3: Adaptive expectations	1.13	1.00

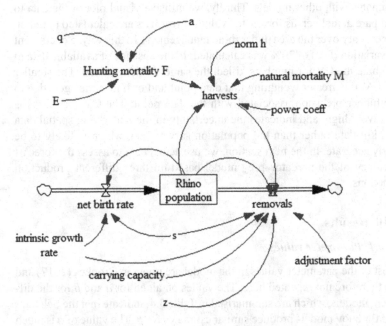

Figure 5.1 Stock–flow diagram for constant effort model (harvesting function model 1).

In spite of this, the model is able to reproduce the historic behaviour of the data extremely well. Projecting the model forward indicates that rhino stocks will recover to carrying capacity (Figure 5.2, top row, first column). Although hunting mortality and CPUE are reasonably well replicated, harvest rates are considerably less than 2012 values.

5.10.3 Model 2: Current stocks, future profitability

The myopic model considers dynamics based on only one system: stocks. The second model relaxes this assumption, and harvesting decisions are a function of both current stocks and expectations of future profitability. The stock–flow diagram for the model in Figure 5.3 indicates that there is now feedback between three components in the models: stocks, poaching effort and profitability. In contrast to the aforementioned myopic model, which was linear, this is a dynamic model. Effort influences both poacher profit as well as stocks, and while stocks influence poacher profit, and profitability in turn influences changes in effort.

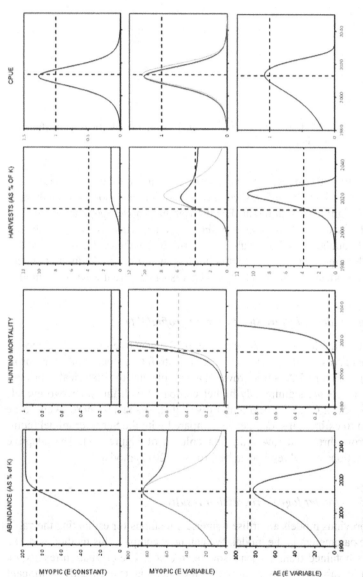

Figure 5.2 Dynamics over time of hunting from different harvesting functions. The intersection of the dashed lines indicates 2012 data. Therefore, values to the right of the vertical dashed lines are all projected values. The black lines are for the Baranov model, while grey lines are for the Schaefer model.

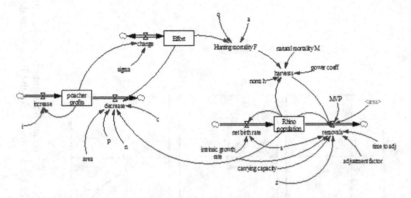

Figure 5.3 Stock–flow diagram for variable effort models (harvesting function models 2 and 3).

The fit of the model with the historical abundance data is as good as the myopic model (Figure 5.2, second row, first column); however the dynamics are significantly different. For the Baranov model, populations decline but stabilise at a long-term equilibrium value, whereas for the Schaefer model populations with variable effort, rhino abundance declines to extinction (Figure 5.2, third row, first column). The difference is due to the density-dependent term β, which affects harvests in different ways in the model.

5.10.4 Model 3: Future stocks, future profitability

The third model simulates the effects of the adaptive expectations model, where harvesting decisions are based on both future stocks and future profitability (Figure 5.2, bottom row). It also replicates the historical data well, but harvests are significantly higher compared with the other two models (Figure 5.2). Once expectations fully adjust, stocks are more likely to be driven to extinction sooner, even compared with the Schaefer model (compare row three and row four, first column of Figure 5.2). The adaptive expectations model replicates the Crookes (2017) model.

5.10.5 Schaefer logistic regression results

The previous models all utilise numerical methods for estimating the value of the parameters in the model. We can now compare this method with the results obtained from the least-squares method. The estimation results are given in Table 5.5. These show that the coefficients are highly significant

Table 5.5 Least-squares regression results for Schaefer logistic specification

	Coefficient	*T stat*	*Statistic/data*	*Sig*
k	0.1627	14.11		***
qk/r	14.9901	5.47		***
Model F			29.96	***
Adj R^2			0.557	
N			24	

Note: *** Significant at least at the 1% level.

Table 5.6 Calculations of MAPE (%) for the three production functions (2010–2015)

Production function	*MAPE (%)*
LV: Cobb–Douglas ($a = 0$; $\alpha = 0.002$; $\beta = 0$)	8.76
LV: Cobb–Douglas ($a = 1$; $\alpha = 0.002$; $\beta = 0$)	7.95
LV: Schaefer ($a = 1$; $\alpha = 1$; $\beta = 1$)	5.55
LV: Baranov ($a = 1$; $\alpha = 0.002$; $\beta = 0$)	8.80
Econ: Schaefer logistic	101.75

Source: Own calculations.

Notes: LV, Lotka–Volterra; Econ, econometric. Cobb–Douglas harvest function given in Equation 5.3.

($p < 0.01$) and the model fit reasonable (adj. $R^2 = 0.557$). However, although the estimate for carrying capacity k is significant, it differs markedly from estimates from the LV model. Here $k = 0.16$ individuals/km², while the LV model using numerical methods to estimate produced an estimate for $k = 0.4$ individuals/km² (Crookes 2017). Clark (1990) notes that bioeconomic models are often sensitive to changes in their underlying parameters. This suggests that forecasts of rhino abundance could differ markedly depending on the parameters employed. It is therefore necessary to assess forecast accuracy using a quantitative measure. Next, we will consider the results of the MAPE calculations.

5.10.6 MAPE calculation

Table 5.6 indicates that the LV models calibrated using numerical methods produced "highly accurate" calibration estimates of the historical data (based on MAPE). This suggests that all three models could be used to make predictions of rhino abundance. By contrast, the LV model based on the least-squares methodology provided a forecast that is "poor". Although a limitation of the MAPE measure is that it overstates the error found in population forecasts (Tayman and Swanson 1999), the measure nonetheless indicates large discrepancies between the LV models based on numerical method calibration and those using the least-squares method. More advanced econometric specifications are possible that may address some of the problems with the traditional Schaefer logistic model (e.g. Zhang and Smith 2011); however this is beyond the scope of this study. The best model, according to the MAPE calculation, was the Schaefer LV model (which is the Cobb–Douglas with adaptive expectations model), estimated using numerical methods [which is also the model used by Crookes (2017)].

In the next section, we consider some of the implications of this for conservation by comparing forecasts of the model forward from 2015 to 2020.

5.10.11 Forecast accuracy

The Schaefer LV model provided the best calibration of rhino abundance for the historical data to 2015 based on the MAPE measure. We compare forecasts from 2015 to 2020 through a visual plot of the data, as well as by comparing the forecasts of the Schaefer LV model with forecasts from Emslie and Adcock (2016).

Figure 5.4 summarises the results of the visual plots for both rhino abundance (Figure 5.4, top graph) and the number killed (Figure 5.4, bottom graph). The results show that the data provide a reasonable fit with the historical data given the uncertainty associated with the data. Also, the forecasts are highly accurate over the short to medium term (two–three years), but that the forecasts are still reasonably accurate over the seven years of the forecast, particularly as it pertains to rhino abundance.

Table 5.7 summarises the results of the comparison of the Schaefer LV model (for black and white rhinos) with forecasts from Emslie and Adcock (2016, white rhinos only). The table indicates that the forecasts by Emslie and Adcock (2016) and Crookes (2017) are remarkably similar, but both underestimate the actual number (in 2019). A reason for this is that neither model could anticipate the effects of the droughts that occurred in 2018–19 on rhino abundance. In spite of this, the forecast capability of both models is highly accurate. Given that Crookes's (2017) model is based on 2012 rhino abundance data, it means that this model has a forecast accuracy of at least seven years.

5.11 Discussion

Crookes and Blignaut (2016) compared the forecasting capabilities of a simple LV system dynamics model modelling intersectoral dynamics with a forecast generated by artificial neural networks (ANNs). They demonstrated

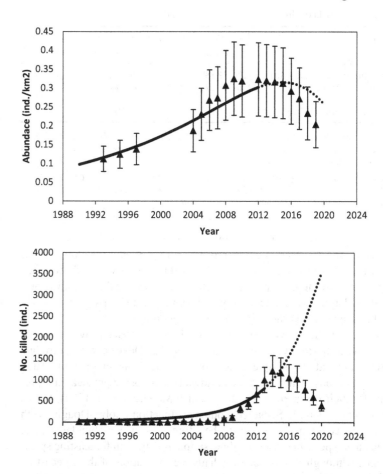

Figure 5.4 Visual plot of (A) rhino abundance (top graph) and (B) number killed (bottom graph). The error bars represent coefficients of variation (CVs) and the triangles actual data. The solid line represents model simulations from the LV Schaefer model ($a = 1$), and the dotted line represents forecasts from the same model.

that these simple LV models based on the logistic model provide a comparable forecast to neural networks over a ten-year period. More recently Li et al. (2017) found that the LV model provided "highly accurate" forecasts of battery electric vehicle (BEV) demand in China. In the present study, we assess the forecast accuracy of the LV model using numerical methods to estimate the value of the parameters (Markov chain Monte Carlo). We compare three LV models based on different production functions and

Table 5.7 Forecasts of rhino abundance, 2015–2020

Forecasts	Year	White rhino[a]	Annual % change	White and black rhino[b]	Annual % change
Starting number (actual data)	2015	18489		20306	
Based on last 5+ years poaching data	2020	16277	–2.5%	16743	–3.8%
Actual population (2019)[c]				13206	

Sources: [a] White rhino estimates from Emslie and Adcock (2016). [b] White and black rhino estimates from Crookes (2017). [c] Actual data from annual reports from Department of Agriculture, Forestry and Fisheries (DAFF).

Notes: Emslie and Adcock (2016) use a growth rate of 0.077 for white rhinos. Crookes (2017) models a growth rate of both black and white rhinos of 0.061. Assumes 100% detection rate. Crookes (2017) estimates based on the Schaefer LV model.

assess the prediction accuracy highly accurate over the short to medium term using MAPE. This is compared with an LV model developed using the least-squares method. The results showed that the LV model estimated using numerical methods produced better estimates for the unknown biological and harvest parameters compared with estimates derived from an LV model estimated using the least-squares methodology.

Previous studies have found that the logistic model provides robust predictions in a variety of sectors. For example, Devezas and Corredine (2001) found that "the simple logistic often outperforms more complicated parameterizations, which have the disadvantage of losing physical interpretations for their parameters" (p. 28; see also Marchetti et al. 1996). In this assessment, the simple Schaefer production function (Cobb–Douglas with adaptive expectations) provided the best forecast of rhino abundance data. This again supports the assertion that "simple is better" in forecasting specifications. Although this is not an exhaustive evaluation of the forecasting capability of the predator–prey system, and there may be instances when other functional forms are preferable, it nonetheless indicates the potential of even these models to provide forecasts of different entities.

The modelling exercise also indicated that just because a model is validated according to statistical criteria, it does not mean that it is suitable for forecasting. The time series regression model provided statistically significant parameter estimates (at least at the 1% level), yet the forecasting capabilities of the model were poor. Our results show that model validation should include an assessment of forecast accuracy by comparing model estimates with a segment of the historical data before it is used to forecast into the realm of the unknown. While forecasting remains a highly imprecise science, with many unknown factors and variables, these validation methods can improve the robustness of predictions.

Our study also sheds light on hunter behaviour. The adaptive expectations model was indicated as the best fit with the historical data. It shows that future stocks are the basis on which poachers form expectations of harvests and that elasticities of substitution of stocks and effort are unity (α, $\beta = 1$, Schaefer model). This means that harvests have fully adjusted to expectations around stocks and are thus higher than would be the case under a Cobb–Douglas specification (with α, $\beta < 1$). This was the most aggressive harvesting regime of the models considered. Poachers do appear to have revised their poaching behaviour downward as a result of declines in rhino abundance, which shows promise that rhino populations may rebound in the future.

References

Armstrong, J.S., 2001. Evaluating forecasting methods. In J. Scott Armstrong (Ed.), *Principles of Forecasting: A Handbook for Researchers and Practitioners*. Boston, MA: Springer.

Cacho, O.J., 1997. Systems modelling and bioeconomic modelling in aquaculture. *Aquaculture Economics & Management*, 1(1–2), pp. 45–64.

Castro, L.M., Härtl, F., Ochoa, S., Calvas, B., Izquierdo, L., and Knoke, T., 2018. Integrated bio-economic models as tools to support land-use decision making: A review of potential and limitations. *Journal of Bioeconomics*, 20(2), pp.183–211.

Clark, C.W., 1990. *Mathematical Bioeconomics: The Optimal Management of Renewable Resources* (2nd edn). New York: John Wiley.

Crookes, D.J., 2017. Does a reduction in the price of rhino horn prevent poaching? *Journal for Nature Conservation*, 39, pp.73–82.

Crookes, D.J. and Blignaut, J.N., 2016. Predator-prey analysis using system dynamics: An application to the steel industry. *South African Journal of Economic and Management Sciences*, 19(5), pp.733–746.

Devezas, T.C. and Corredine, J.T., 2001. The biological determinants of long-wave behavior in socioeconomic growth and development. *Technological Forecasting and Social Change*, 68(1), pp.1–57.

Emslie, R. and Adcock, K., 2016. A conservation assessment of *Ceratotherium simum simum*. In M.F. Child, L. Roxburgh, E. Do Linh San, D. Raimondo, and H.T. Davies-Mostert (Eds), *The Red List of Mammals of South Africa, Swaziland and Lesotho*. South Africa: South African National Biodiversity Institute and Endangered Wildlife Trust.

Grimsrud, K.M., Chermak, J.M., Hansen, J., Thacher, J.A., and Krause, K., 2008. A two-agent dynamic model with an invasive weed diffusion externality: An application to Yellow Starthistle (*Centaurea solstitialis* L.) in New Mexico. *Journal of Environmental Management*, 89(4), pp.322–335.

Gujarati, D., 2003. *Basic Econometrics* (4th edn). New York: McGraw-Hill.

Hung, H.C., Tsai, Y.S., and Wu, M.C., 2014. A modified Lotka–Volterra model for competition forecasting in Taiwan's retail industry. *Computers & Industrial Engineering*, 77, pp.70–79.

Hung, H.C., Chiu, Y.C., Huang, H.C., and Wu, M.C., 2017. An enhanced application of Lotka–Volterra model to forecast the sales of two competing retail formats. *Computers & Industrial Engineering*, 109, pp.325–334.

Jones, R., Cacho, O., and Sinden, J., 2006. The importance of seasonal variability and tactical responses to risk on estimating the economic benefits of integrated weed management. *Agricultural Economics*, 35(3), pp.245–256.

Lee, S.J., Lee, D.J., and Oh, H.S., 2005. Technological forecasting at the Korean stock market: A dynamic competition analysis using Lotka–Volterra model. *Technological Forecasting and Social Change*, 72(8), pp.1044–1057.

Li, S., Chen, H., and Guofang Zhang, G., 2017. Comparison of the short-term forecasting accuracy on battery electric vehicle between modified Bass and Lotka-Volterra model: A case study of China. *Journal of Advanced Transportation*, 2017, Article ID 7801837.

Liu, X. and Heino, M., 2014. Overlooked biological and economic implications of within-season fishery dynamics. *Canadian Journal of Fisheries and Aquatic Science*, 71, pp.181–188.

Marasco, A., Picucci, A., and Romano, A., 2016. Market share dynamics using Lotka–Volterra models. *Technological Forecasting and Social Change*, 105, pp.49–62.

Marchetti, C., Meyer, P.S., and Ausubel, J.H., 1996. Human population dynamics revisited with the logistic model: How much can be modeled and predicted? *Technological Forecasting & Social Change*, 52, pp.1–30.

MarketsandMarkets, 2020. *Weather Forecasting Services Market by Industry (Aviation, Agriculture, Marine, Oil & Gas, Energy & Utilities, Insurance, Retail, Media), Forecasting Type (Nowcast, Short, Medium, Long), Purpose, Organization Size & Region - Global Forecast to 2025*. Northbrook, IL: MarketsandMarkets.

McDermott, S.M., Irwin, R.E., and Taylor, B.W., 2013. Using economic instruments to develop effective management of invasive species: Insights from a bioeconomic model. *Ecological Applications*, 23(5), pp.1086–1100.

Mentzer, J.T. and Kahn, K.B., 1995. Forecasting technique familiarity, satisfaction, usage, and application. *Journal of Forecasting*, 14, pp.465–476.

Paul, C., Reith, E., Salecker, J., and Knoke, T., 2019. How integrated ecological-economic modelling can inform landscape pattern in forest agroecosystems. *Current Landscape Ecology Reports*, 4(4), pp.125–138.

Pella, J.J. and Tomlinson, P.K., 1969. A generalised stock-production model. *Bulletin of the Intertropical Tuna Commission*, 13, pp.421–458.

Ponce-Marbán, D., Hernández, J.M., and Gasca-Leyva, E., 2006. Simulating the economic viability of Nile tilapia and Australian redclaw crayfish polyculture in Yucatan, Mexico. *Aquaculture*, 261(1), pp.151–159.

Sherden, W.A., 1998. *The Fortune Sellers*. New York: Wiley.

Tayman, J. and Swanson, D.A., 1999. On the validity of MAPE as a measure of population forecast accuracy. *Population Research and Policy Review*, 18, pp.299–322.

The Telegraph, 2011. Trend-spotting is the new £36bn growth business. 01 May 2011.

WSGN, 2021. Insights. London, England: World's Global Style Network. Available at: https://www.wgsn.com/en/ (accessed on 22 February 2021).

Wu, L. and Liu, S., 2013. Using grey Lotka-Volterra model to analyze the relationship between the gross domestic products and the foreign direct investment of Ningbo city. In *Proceedings of 2013 IEEE International Conference on Grey systems and Intelligent Services (GSIS)*, Macao, China, pp.265–268, doi: 10.1109/GSIS.2013.6714796.

Zhang, J. and Smith, M.D., 2011. Estimation of a generalized fishery model: A two-stage approach. *Review of Economics and Statistics*, 93(2), pp.690–699.

6 Co-evolutionary models and the prisoner's dilemma game

6.1 Introduction

In Chapter 2 we saw that there are three types of property regimes: (1) pure open access, (2) private property and (3) open access with "rules" (or communalism). The open-access stock equilibrium level is below that of the sole owner (private property) and open access with rules' stock level is typically between the private property and sole ownership level.

Here, we consider how these different property rights regimes play out in the contexts of resource scarcity and resource abundance, and under what conditions it is optimal to co-operate and under what conditions it is optimal to "cheat". In order to do this, we present four co-evolutionary models, one for each of the elements of a modified prisoner's dilemma game, and then use the equilibrium outcomes from these co-evolutionary models to determine the payoffs for this game.

In the next section, the prisoner's dilemma game is presented, followed by the four bioeconomic models that are used to determine the payoffs. Finally, the revised prisoner's dilemma game is presented, and the implication thereof is considered.

6.2 The prisoner's dilemma game

The prisoner's dilemma game hypothesises that two prisoners held in separate cells are given the opportunity to either co-operate with each other or defect (turn state's evidence against the other prisoner; Table 6.1). If only one prisoner defects, that one gets released but the other one gets a longer sentence. If both defect, however, they remain incarcerated. If neither defects, they get a shorter sentence. The prisoner's dilemma game states that it is optimal to co-operate (mutualism payoff = 3 for both players) but that greed can prevent mutualism (if one defects that one gets a payoff of 5 compared with the other who gets 0, but if both defect they get a punishment of 1 each).

DOI: 10.4324/9781003247982-6

Table 6.1 Prisoner's dilemma game

		Column player	
		Co-operate	Defect
Row player	Co-operate	R = 3, R = 3 Reward for mutual cooperation	S = 0, T = 5 Sucker's payoff, and temptation to defect
	Defect	T = 5, S = 0 Temptation to defect and sucker's payoff	P = 1, P = 1 Punishment for mutual defection

Source: Axelrod and Hamilton (1981).

Note: The payoffs to the row chooser are listed first.

The game shows that the optimal strategy (the Nash equilibrium) is for both players to defect.

In the next section we propose four game-theoretic outcomes based on interactions between two players.

6.3 Four game-theoretic outcomes

In order to determine the payoffs in our modified prisoner's dilemma game, we first consider four game-theoretic outcomes based on a two-player game (Table 6.2).

In the next section, we present the bioeconomic framework.

Table 6.2 Four game-theoretic outcomes

		Player 1	
		Resource abundance	Resource scarcity
Player 2	Abundance	Utopia – A system characterised by abundant resources and harmony between different species, working together and not in conflict	Survival of the fittest – One dominant entity with the other entities subservient (this is the private property outcome when one resource is scarce)
	Scarcity	Survival of the fittest – One dominant entity with the other entities subservient (this is the private property outcome when one resource is scarce)	Commons, but is it a tragedy? There is no dominant entity, but there is substitution between different entities in the context of resource scarcity

6.4 Bioeconomic framework

6.4.1 Generalised system

In order to determine the payoffs for this modified prisoner's dilemma game, we utilise a generalised two-variable predator–prey formulation:

$$F(x,y) = rx\left(1 - \left(\frac{x}{K}\right)^m\right) + \alpha xy \qquad (6.1)$$

$$G(x,y) = sy\left(1 - \left(\frac{y}{L}\right)^n\right) + \beta xy \qquad (6.2)$$

where $F(x,y)$ presents the evolution of player 1, with x representing the stock of player 1 with growth rate r and carrying capacity K, that converts resources at rate α. Similarly, $G(x,y)$ represents the player 2 system, with y representing some stock of player 2 with growth rate s and carrying capacity L, that converts resources at rate β. The parameters m and n are the skewness parameters previously given (see Equation 2.1 and associated text).

The player could be a biological entity or an economic entity. For example, the work of emeritus professor Lim Ching Yah demonstrates that, for many economies, an S-shaped growth curve is indicative (see Crookes and Blignaut 2015 for an elaboration).

6.4.2 Resource abundance

The four models are differentiated as follows (resource abundance scenarios):

1. Utopia. This is the prisoner's dilemma case where both co-operate (top right quadrant in Table 6.2). Mathematically, it occurs when $\alpha > 0$ and $\beta > 0$. With $\alpha > 0$ it means that player 2 enhances the growth of player 1. When $\beta > 0$, player 1 enhances the growth of player 2. This implies that there is a growth of both player 1 and player 2.
2. Survival of the fittest (evolutionary model). In the prisoner's dilemma game, this occurs when player 1 defects and player 2 co-operates (top left quadrant in Table 6.2). This is the predator–prey model, with $\alpha > 0$ and $\beta < 0$, and means that player 2 acts as a sink for player 1. The player 2 stock availability limits the future expansion of player 1 ($\alpha < 0$). $\beta < 0$ means that player 1 consumes player 2. When $\alpha > 0$, this means that player 1 grows by consuming player 2.
3. Survival of the fittest (evolutionary model). This is the bottom right quadrant in Table 6.2 and occurs when player 2 defects and player 1 co-operates. This occurs mathematically when $\alpha < 0$ and $\beta > 0$. Player

2 consumes player 1 ($\alpha < 0$), but player 1 enhances the growth of player 2 ($\beta > 0$).

4. Tragedy of the commons. If both parties defect, they get penalised.

Commons, but is it a tragedy? The scenarios are re-run, but this time with the Gause model (with $\alpha < 0$ and $\beta < 0$, but with $\alpha = \beta = r = s$, $K = L = 1$, and $n = m$ in the case of co-operation). Player 2 consumes player 1 at the same rate as player 1 consumes player 2. This is co-operation under resource scarcity.

6.5 Modified prisoner's dilemma game

Revisiting our modified prisoner's dilemma game in the context of resource abundance, it can be shown that the payoffs from the bioeconomic models in the previous section are shown in Table 6.3. It is evident that the Nash equilibrium is found under the scenario where both player 1 and player 2 co-operate with each other. This is the utopia scenario for the co-evolutionary model. The temptation to defect (and take payoff K or L) actually constitutes a suboptimal outcome in this scenario.

However, in most cases resource scarcity is a reality for many systems. Table 6.4 provides the modified prisoner's dilemma game in the context of resource scarcity. In this case, the Nash equilibrium is actually to defect because payoff K or L is greater than the incentive to co-operate. If both players defect, there is a penalty. But, if both players co-operate then the payoffs are actually greater than the maximum sustainable yield (MSY) (in the case of long-lived species). The exercises at the end of the chapter provide a means of proving this).

Table 6.3 Modified prisoner's dilemma game with payoffs from co-evolutionary models (resource abundance scenario)

		Column player (x)	
		Co-operate	Defect
Row player (y)	Co-operate	$R = \infty$, $R = \infty$ Reward for mutual co-operation	$S = 0$, $T = K$ Sucker's payoff, and temptation to defect
	Defect	$T = L$, $S = 0$ Temptation to defect and sucker's payoff	$P = 1$, $P = 1$ Penalty for defection

Note: The payoffs to the row chooser are listed first. K and L are the carrying capacity of x and y, and m and n are the skewness parameters of the Pella and Tomlinson equations. See Equations 6.1 and 6.2 for specification of payoffs.

Table 6.4 Modified prisoner's dilemma game with payoffs from co-evolutionary models (resource scarcity scenario)

		Column player *(x)*	
		Co-operate	Defect
Row player *(y)*	Co-operate	$R = (1-x)^{\frac{1}{n}} \cdot L,\ R = (1-y)^{\frac{1}{m}} \cdot K$ Reward for mutual cooperation	$S = 0,\ T = K$ Sucker's payoff, and temptation to defect
	Defect	$T = L,\ S = 0$ Temptation to defect and sucker's payoff	$P = 1,\ P = 1$ Penalty for defection

Note: The payoffs to the row chooser are listed first. K and L are the carrying capacity of x and y, and m and n are the skewness parameters of the Pella and Tomlinson equations. See Equations 6.1 and 6.2 for the specification of payoffs.

6.6 Discussion

Bioeconomic models have had few applications in the context of mutualism (Law 1985; Magain et al. 2017). In this chapter we find that mutualistic models are most effective for promoting economic growth and well-being. These results are not new. Almost three decades ago, May (1982) showed that beneficial mutualism exhibits an exponential increase in population for both species involved in the mutualistic relationship. Here, we propose using co-evolutionary models to inform conditions for co-operation and defection under resource abundance and resource scarcity. We find that the modified prisoner's dilemma game, where the Nash equilibrium is subject to the mutualism (utopia) outcome, is optimal under resource abundance. However, this outcome is not realistic for most natural resource systems. Indeed, the economics problem is how to allocate scarce resources (Conrad 1999).

Under resource scarcity, mutualism is not the Nash equilibrium but does provide payoffs that exceed MSY (in the case of long-lived species). This means that co-operation will not automatically emerge in these systems but needs to be regulated and enforced. Under resource harvesting with rules (the communalism outcome), gains can be in excess of MSY (and therefore comparable to the private property outcome) but are still less than if one of the players defects. This probably explains why many communal areas are degraded. Incentives should therefore be provided for communities to work together, to ensure win–win outcomes.

Exercises

1. The case where $m = n = 1$. Use Equations 6.1 and 6.2, and insert the following parameter values:

	Utopia	Predator–prey	Predator–prey	Commons
Growth rate (r, s)	0.1	0.1	0.1	0.1
Carrying capacity (K, L)	1.0	1.0	1.0	1.0
Alpha	0.1	0.1	−0.1	−0.1
Beta	0.1	−0.1	0.1	−0.1

Show that the payoffs for players 1 and 2 increase exponentially under the utopia scenario, increase to carrying capacity when one of the players' defects (and the payoffs for the other player decreases to zero) and the payoffs for both players under the resource scarcity (where both players co-operate) converge on MSY. (Hint: MSY = $0.5K$ in the logistic model.)

2. By setting Equations 6.1 and 6.2 equal to zero, show that

$$xss = \left(\frac{r - \alpha y}{r}\right)^{\frac{1}{m}} K, 0 < y < 1$$

and

$$yss = \left(\frac{s - \beta x}{s}\right)^{\frac{1}{n}} L, 0 < x < 1$$

3. Show that, under conditions of co-operation, $xss = yss < msy$ for $n = m < 1$, $xss = yss = msy$ for $n = m = 1$, and $xss = yss > msy$ for $n = m > 1$. (Hint: If $r = s = \alpha = \beta$, and $K = L$, then $x = y$. Also, see Chapter 2, footnote 1 for the equation for MSY under the Pella and Tomlinson model.)

References

Axelrod, R. and Hamilton, W.D., 1981. The evolution of cooperation. *Science*, 211(4489), pp.1390–1396.

Conrad, J.M., 1999. *Resource Economics*. New York: Cambridge University Press.

Crookes, D.J. and Blignaut, J.N., 2015. Debunking the myth that a legal trade will solve the rhino horn crisis: A system dynamics model for market demand. *Journal for Nature Conservation*, 28, pp.11–18.

Law, R., 1985. Evolution in a mutualistic environment. In D.H. Boucher (Ed.), *The Biology of Mutualism: Ecology and Evolution*. Oxford: Oxford University Press, pp.145–170.

Magain, N., Miadlikowska, J., Goffinet, B., Sérusiaux, E., and Lutzoni, F., 2017. Macroevolution of specificity in cyanolichens of the genus Peltigera section Polydactylon (Lecanoromycetes, Ascomycota). *Systematic Biology*, 66(1), pp.74–99.

May, R.M., 1982. Mutualistic interactions among species. *Nature*, 296(5860), pp.803–804.

7 Co-evolutionary models and oceans governance

The case of the African penguin (*Spheniscus demersus*)

7.1 Introduction

The small pelagic fishing sector in South Africa is facing restrictions on sardine (*Sardinops sagax*) and anchovy (*Engraulis capensis*) netting as part of a new government plan to save endangered African penguins (*Spheniscus demersus*) from starvation and ultimate extinction (Carnie 2019). The question arises, what is the optimal management strategy for these species?

Here, we develop a predator–prey model for oceans governance, using a co-evolutionary model with the African penguin as the predator, and anchovy and sardine as the prey. Economic returns arising from the natural resource stock (anchovy and sardine fishing) is exogenous to the predator–prey model, and is estimated using the total revenues from these fisheries, revenues from the wholesale sector (canning and associated industries), and then the multiplier effects on the economy as a whole from the fishing industry is incorporated. This is compared with the non-market values associated with African penguin viewing. The effect on the economy as a whole of these tourism values is also then modelled using input–output multiplier analysis (which is a conservative estimate of the economy-wide impact of tourism expenditure).

The model therefore includes both marketed and non-marketed ecosystem values. The approach is similar to Conrad (2019), who developed a dynamic optimisation model that also considered one marketed and one non-marketed ecosystem service for oysters in the Chesapeake Bay. His approach used water quality and oyster stocks as the state variables, whereas here we use the predator and prey species as the state variables.

Weller et al. (2014) developed a penguin model, focussing specifically on the Robben Island region. Theirs was an ecological model, with a more elaborate population component than ours; however, they did not explore economic linkages in their paper. They found that penguin populations were "strongly driven by food availability and to a lesser degree by oiling and

DOI: 10.4324/9781003247982-7

marine predation, while climate events and terrestrial predation had low impacts".

Robinson et al. (2015) developed a Bayesian model to estimate penguin population parameters for the Robben Island region. They found, first, that penguins prefer sardines to anchovies due to the higher calorific content of sardines. Their model therefore only incorporates sardines. Second, Robinson et al. found that natural mortality due to changes in the availability of prey species was the main determinant.

Our model is developed for the period 1987–2018 (historical time period) in order to determine which sectors are most profitable. The model is then forecast forward from 2020 in order to consider the effects on stocks and economic welfare of different management scenarios, namely closing the fishery or keeping the fishery open. The effect of environmental stochasticity is also taken into consideration.

7.2 Methodology

7.2.1 Predator–prey model

The population of the predator in time t, African penguins (Y_t), evolves according to the following dynamic equation:

$$Y_{t+1} = Y_t + n_1 \left[\theta \left(\gamma^1 X_t^1 + \gamma^2 X_t^2 \right) r_1 Y_t \left(1 - \frac{Y_t}{K_1} \right) - m Y_t \right]$$

where n_1 is an adjustment coefficient, r_1 is the intrinsic growth, K_1 is the carrying capacity of African penguins and m is the mortality coefficient. The coefficient θ estimates the effect of diet (the prey species) on the growth rate of the predator (African penguins), γ^1 is the proportion of the diet that comes from prey species 1 (sardines) and γ^2 is the proportion of the diet that comes from prey species 2 (anchovies). The biomass variable X_t^1 is the stock of sardines in time t and X_t^2 is the stock of anchovies in time t.

The population of the prey species P (sardines, anchovies) in time t (X_t) evolves according to the following equation:

$$X_{t+1}^P = X_t^P + n_2^P \left[r_2^P X_t^P \left(1 - \frac{X_t^P}{K_2^P} \right) - w^P X_t^P Y_t \right]$$

where n_2^P is an adjustment coefficient for prey species P, r_2^P is the intrinsic growth rate, K_2^P is the carrying capacity of the prey species P (anchovies, sardines) and w^P is the mortality coefficient associated with fishing and natural predation. Fishing harvests of the prey species h_2^P equals

$$h_t^P = u^P X_t^P Y_t$$

And mortality due to penguin interactions equals

$$m_t^P = z^P X_t^P Y_t$$

Therefore

$$w^P = u^P + z^P$$

where u^P is fishing mortality and z^P is natural mortality.

7.2.2 Economy-wide impacts

The following equations were used to estimate the economy-wide impacts of penguins:

$$\pi_t^Y = tm \; x \; p_1 Y_t$$

where π_t^Y is the economy-wide impact of penguins, tm is the tourism multiplier of penguins, p_1 is the non-market (viewing) value of penguins, ρ is the discount factor $\left(\dfrac{1}{1+i}\right)$ and i is the discount rate.

7.3 Data and model estimation

7.3.1 Data sources

These models are relatively data intensive. It is therefore necessary to use a number of different techniques to obtain the values of the parameters. Three are mentioned here briefly. The first is obtaining the value of the parameters from published literature sources. The second is to estimate the value of the parameters from regression models. The third is to use calibration to obtain values for the missing parameters in the model. This can only be done if historical time series for the stock variables (predator, prey abundance) is available. All three of these techniques were used in the current model.

Data is utilised for the entire population of penguins and prey (sardines, anchovies) present in South Africa at the time of the sample for a given year. Time series for anchovy and sardine biomass are obtained from the Department of Agriculture, Forestry and Fisheries (DAFF 2017), and for African penguins from the Department of Environmental Affairs (DEA 2019). The prey data was available for the entire sample (from 1987 to 2017) and is expressed in tonnes. The predator data was available from 1999 to 2018 and is expressed in breeding pairs. We used data from Waller (2011) for six sites in South Africa (Dassen Island, Robben Island, Boulders, Stoney Point, Dyer Island and St Croix Island) and the implied ratio of breeding pairs (total) to breeding pairs at these six sites in 1999 to estimate the total number of breeding pairs in South Africa between 1987 and 1998. The total number of data points for the sample is therefore 31. This was then benchmarked against the estimate of breeding pairs in South Africa in 1978–1979 of 70,000 to ensure that overall trends were consistent (Birdlife International 2019). Figure 7.1 indicates the historical population trends of these three species.

7.3.2 System dynamics models

Calibration of predator–prey models is best achieved using system dynamics modelling software such as Vensim (Crookes and Blignaut 2019). A number of other tools are also available in Vensim in order to achieve the stated goal of the model, namely the estimation of the optimal time path of the stock variables. The model was therefore constructed in Vensim.

7.3.3 Regression analysis

A multivariate regression analysis was conducted in order to estimate the trophic coefficient θ, in other words the proportion of prey species γ^1 and γ^2 in the predator's diet. It is immediately evident that anchovies are not a

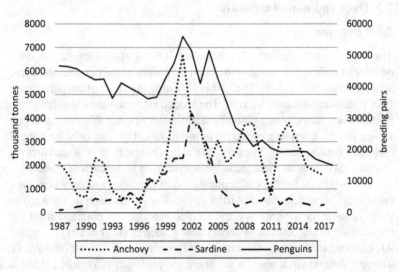

Figure 7.1 Number of breeding pairs of African penguins (right axis), and sardine and anchovy biomass (left axis) in 1987–2017 in South Africa.

significant explanatory variable of penguin diet on a national level. Only sardines are a significant explanatory variable of the penguin diet. Based on the regression analysis (Table 7.1), a value of 0.0059 was assigned to θ, $\gamma^1 = 1$ and anchovy was excluded from the model ($\gamma^2 = 0$). The final model was thus a two-species predator–prey model (one predator = African penguins, and one prey = sardines).

7.3.4 Stock–flow diagram

The construction of the model in Vensim enables a visual display of the interaction among the different parameters in the model (Ford 2009). This is called a stock–flow diagram. The stock–flow diagram for the present two-species predator–prey system is displayed in Figure 7.2.

7.3.5 Parameters used in the model

The parameters used in the model, and their sources, are reported in Table 7.2.

7.3.6 Stochastic component

The historical data shows a spike in recruitment for sardines between the years 2001 and 2007. Although the exact cause of this spike in recruitment is unknown and is regarded as "environmental", this spike in recruitment does appear to occur at intervals of roughly 20 years (De Moor, personal communication). These decadal oscillations in recruitment are a worldwide phenomenon among small pelagics (e.g. Chavez 2003; Bertrand et al. 2004; Lindegren et al. 2013; Champagnon et al. 2018). We therefore add a stochastic component to the model by simulating an exogenous environmental shock to sardine recruitment every 20 years in order to take into consideration these environmental variations. The results of this on the dynamics of penguin and sardine abundance are given in Figure 7.3. The results show an improved (although not perfect) correlation with the historical data.

Table 7.1 Multivariate regression analysis of predator–prey interaction terms

	Intercept	Sig	Anchovy	Sig	Sardines	Sig	F model	Sig	Adj R^2	n
Model 1	31445	***	−0.0013	ns	0.0067	***	5.17	**	0.217	31
Model 2	33147	***	0.0009	ns			0.32	ns	−0.023	31
Model 3	29340	***			0.0059	***	9.74	***	0.226	31

Notes: Dependent variable = number of penguin breeding pairs; Sig = significance; *** significant at the 1 percent level; ** significant at the 5 percent level; ns = not significant; n = number of observations; Adj R^2 = adjusted R^2.

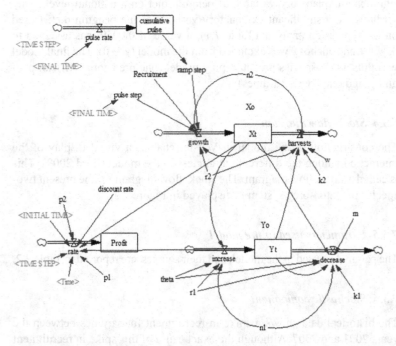

Figure 7.2 Stock–flow diagram for the predator–prey system.

7.3.7 Model evaluation

The model provides a reasonable (although not good) fit with the historical data (based on estimates of mean absolute percentage error). The visual plot of the data (Figure 7.3) shows that the model does not provide a good predator of short-term fluctuations in penguin and sardine populations, and therefore should not be used for short-term (tactical) advice. However, long-term trends are captured well and therefore should rather be used for strategic advice (5–10 years).

Results are discussed in the next section.

7.4 Results

7.4.1 Steady-state values

By setting $Y_{t+1} = Y_t$, the steady-state value of the penguin population Y^* can be obtained. Similarly, by setting $X_{t+1} = X_t$, the steady-state value of the sardine population X^* can be obtained. The equations for these two steady-state values are

Table 7.2 Parameters used in the model (Dmnl = dimensionless)

Parameter	Description	Value	Unit	Source
p2	Price (including multiplier) of sardines	2140	$/tonne	Calculation (see Appendix)
k1	Carrying capacity penguins	500000	Bp	Assumed[1]
Theta	Effect of sardines on penguins	0.0059	Dmnl	Regression model (Table 7.1)
m	Mortality coefficient penguins	0.06	Dmnl	Estimated through calibration
r1	Intrinsic growth rate penguins	0.387	Dmnl	Petersen et al. (2006)
n2	Adjustment parameter sardines	1	1/Year	Assumed
Yo	Number of penguin breeding pairs in 1987	46644	Bp	DEA (2019)
Xo	Biomass of sardines in 1987	111000	tonne	DAFF (2017)
k2	Carrying capacity sardines	6.00E+06	tonne	Assumed[a]
p1	Price (including multiplier) of penguins	54836	$/bp	Calculation (see Appendix 1)
i	Discount rate	0.08	Dmnl	Conningarth Economists (2014)
w	Mortality coefficient sardines	2.00E-07	1/bp	Calibration
r2	Intrinsic growth rate sardines	0.052	Dmnl	Lo et al. (1995)
n1	Adjustment parameter penguins	1	Dmnl	Assumed
tm	Tourism multiplier	1.9	Dmnl	Saayman et al. (2000)
fm	Fishery multiplier	1.5	Dmnl	Brick and Hasson (2016)

[a] Model not sensitive to this parameter.

$$Y^* = \frac{\theta k_1 X^* r_1}{m k_1 + \theta X^* r_1}$$

and

$$X^* = \frac{r_2 k_2 - w Y^* k_2}{r_2}$$

We can see from this that if we set the penguin mortality coefficient (m) equal to 0, the steady-state value of Y^* converges on the penguin carrying capacity k_1. Similarly, if we set the mortality coefficient of penguins (w) equal to 0, the steady-state population of sardines X^* converges on its carrying capacity (k_2). Substituting this value for X^* into Y^* gives

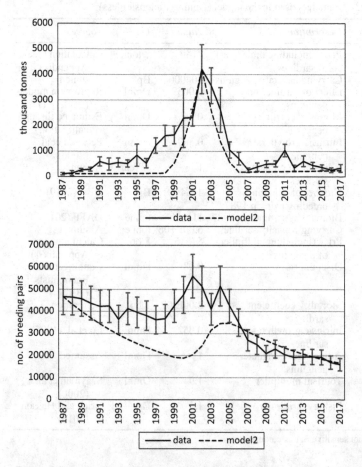

Figure 7.3 Comparisons between historical abundance and model-generated predictions of (A) penguins and (B) sardines. Model 2 = stochastic model. The standard deviation is used for the error bars in panel B, whereas the mean CV for sardines (17%) is used for penguins (panel A) in the absence of other data.

$$Y^* = \frac{\theta k_1 k_2 r_1}{m k_1 + \theta k_2 r_1}$$

Therefore, if there are no harvests of sardines, penguins still do not converge on their carrying capacity since the steady-state value of the penguin population is still dependent on the natural mortality of penguins (assumed exogenous).

7.4.2 Historical value of penguins and sardines (years 1987–2018)

The economy-wide value of penguins in the South African economy is R 11.1 billion per year (USD 0.84 bn, 2017 prices). This is marginally higher than sardines, at R 9.1 billion per year (USD 0.68 bn, 2017 prices). This equates to an economy-wide value from penguins and sardines of USD 25.61 billion at 2018 prices net present value (NPV).

7.4.3 Forecasts: Year 2020 onwards

7.4.3.1 Sardine fishery remains open

Forecasting penguin and sardine populations forward from 2020 to perpetuity, and assuming under the baseline that the sardine fishery remains open, the economy-wide value of penguins and sardines is $48.03 bn (net present value), and sardines fluctuate around a mean of roughly 3 million tonnes, and penguins fluctuate around a mean of just over 30,000 breeding pairs (Figure 7.4, top row).

7.4.3.2 Sardine fishery closes

Setting the sardine mortality coefficient (w) equal to zero, one can model the effect of closing the sardine fishery on projected estimates of sardine and penguin populations, and also the economy-wide value of these two

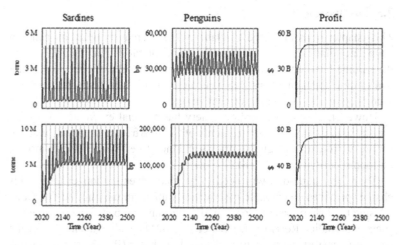

Figure 7.4 Co-evolution of stocks of sardines and penguins and profits, from the year 2020 to perpetuity. (A) Top row shows with fishing (no parameters changed). (B) Second row shows the effect of the closure of the sardine fishery (setting sardine mortality coefficient w equal to zero).

sectors. The NPV (economy-wide benefits) increases to $67.43 bn, and sardine populations increase to around 7.5 million tonnes, and penguin populations increase to around 120,000 breeding pairs.

7.4.6 Sensitivity analysis

We model the sensitivity of the penguin population to changes in the food availability using Monte Carlo simulation, with θ drawn from a random uniform distribution (0;1). An ensemble of 200 realisations was employed. The results indicate that penguin populations are highly sensitive to changes in food availability, particularly in the short term, but even long-term steady-state population numbers range quite substantially depending on the value of θ (Figure 7.5, top left).

Next, we used a similar distribution to estimate the impact of natural mortality m on penguin populations. The results show that penguin populations respond slower to changes in natural mortality and are also not as pronounced as changes in food availability (Figure 7.5, bottom left).

The right column of Figure 7.5 shows the effect on societal economic welfare. The changes in food availability (Figure 7.5, top right) have a greater impact on societal economic welfare compared to changes in natural mortality (Figure 7.5, bottom right).

7.5 Discussion

This chapter finds that food availability has a greater impact on penguin abundance than mortality due to predation or habitat-related effects. Closing the sardine fishery therefore produces the most beneficial effect – penguin and sardine abundance – as well as economy-wide benefits from these two sectors (primarily increased tourism values).

However, closing the sardine fishery has a social externality. Sardines are staple diets among poor segments of South Africa and provide a cheap source of protein, are high in nutrients and therefore have beneficial impacts on food security (Isaacs 2016).

Therefore, trade-offs will need to be made over the longer term. This chapter provides an example case where mutualism may not be desirable. Closing the fishery has economic and environmental benefits, but there is a social externality. Recently, De Wit et al. (2021) developed a model for the economic cost of the polyphagous shot hole borer (PSHB) and *Fusarium* fungus interactions. The PSHB tunnels into certain tree species and enables the spread of the *Fusarium* fungus, which causes trees to die back in certain cases. This has an economic cost on society. Here, there is an environmental

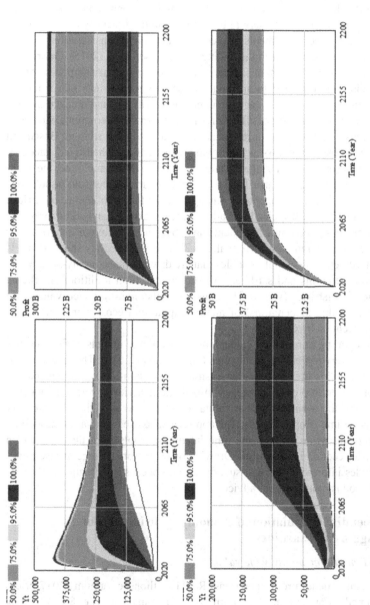

Figure 7.5 Monte Carlo simulation assessing the sensitivity of food availability (top graph) and natural mortality (bottom graph) on penguin population dynamics (left column) and society economic welfare (right column).

mutualism that results in an economic externality. The recommended management measure is to put systems in place to terminate this mutualism.

Similarly, it is highly likely that the sardine population will rebound in the next few years. Therefore, over the longer term, it is not desirable to close the sardine fishery, even though the economic and environmental returns will be lower. This will be necessary to ensure that the food security and livelihood needs of the poorer segments of society are taken into consideration.

Our results concur with the findings of Robinson et al. (2015). Our regression analysis indicates that sardines are a more important component of the penguin diet compared to anchovies. Also, our model finds that if natural mortality can be reduced, penguin populations will rebound quicker, thereby maximising economic rents. These results concur with international studies of penguins. For example, our results are consistent with the study of Busch and Cullen (2009) for yellow-eyed penguin populations, which also indicated that growth is sensitive to changes in adult mortality. At the same time, the claim of Weller et al. (2014) that food availability is driving penguin populations is also supported by our model. Changes in the food availability of African penguins also affect societal welfare to a greater extent than natural mortality.

In the past, economic arguments have been used to favour fisheries over the needs of penguins (Lewis et al. 2012). The present study shows that it is primarily environmental variables that are driving the economy-wide value of both penguins and sardines. We incorporate both traditional marketed values for fisheries (value of landings), but also non-marketed values (the viewing value of penguins) of marine capital and ecosystem services, as recommended by Costanza (1999).

Integrating non-market values into bioeconomic models of fisheries has become very topical in recent years (e.g. Lee et al. 2017). This chapter contributes to this literature by demonstrating how non-market values may be incorporated into co-evolutionary models for fisheries governance, without requiring the assumption that capital be endogenous or that the property rights regime is open access. The approach also uses simulation modelling in Vensim, while at the same time integrating tools from the mainstream fisheries economics literature. Drawing on tools from a range of disciplines provides improved decision support around oceans governance and offers more robust management advice.

Appendix: Calculation of economy-wide impacts of penguins and sardines

A.1 Penguin economy-wide value

Tourism expenditure on penguins: R 311 million per annum (2017 prices assumed) (Van Zyl and Kinghorn 2018). The majority (i.e. 88%) coming from international tourists, 8% from domestic tourists and the remaining 6% from Cape Town residents. Penguin numbers: 2017, 854 breeding pairs (DEA 2019). Therefore, the 2017 economic value of all penguins is R 0.364 million per annum per breeding pair.

The tourism multiplier (domestic) is 1.96 (input–output only) and (foreign) 1.9 (input–output only) (Saayman et al. 2000). The economy-wide value of penguins per breeding pair is R 0.69 million per breeding pair (2017 prices). Converting to 2018 prices using the consumer price index (CPI), and converting to US dollars gives a penguin value of USD 54836 per breeding pair (2018 prices).

A.2 Sardines economy-wide value

The landed value of sardines = R 3856/tonne (2013 price) (Hutchings et al. 2014)

Vessel crew commission = R 767/tonne (2013 price)

Processor's revenue: Canned sardine = R 3820/tonne (2013 price) (80% of catch)

Frozen sardine = R 5000/tonne (2013 price) (20% of catch)

Change in seasonal workers earnings = R 1100/tonne (2013 price) (Hutchings et al. 2014)

Therefore, total price sardine = R 14543/tonne (2013 price)

The fishery multiplier = 1.5 (input–output only) (Brick and Hasson 2016)

Total economy-wide impact of sardine = R 21814/tonne (2013 price). Converting to 2018 prices using the CPI, and converting to US dollars gives USD 2140/tonne (2018 prices). The 2018 R/$ exchange rate is 13.3.

References

Bertrand, A., Segura, M., Gutierrez, M., and Vasquez, L., 2004. From small-scale habitat loopholes to decadal cycles: A habitat-based hypothesis explaining fluctuation in pelagic fish populations off Peru. *Fish and Fisheries*, 5, pp.296–316.

BirdLife International, 2019. Species factsheet: *Spheniscus demersus*. Available at: http://www.birdlife.org (accessed on 7 November 2019).

Brick, K. and Hasson, R., 2016. *Valuing the Socio-economic Contribution of Fisheries and Other Marine Uses in South Africa*. Cape Town: Environmental Economics Policy Research Unit, University of Cape Town.

Busch, J. and Cullen, R., 2009. Effectiveness and cost-effectiveness of yellow-eyed penguin recovery. *Ecological Economics*, 68(3), pp.762–776.

Carnie, T., 2019. Curb mooted to save penguins. *Sunday Times*, 3 November.

Champagnon, J., Lebreton, J.D., Drummond, H., and Anderson, D.J., 2018. Pacific Decadal and El Niño oscillations shape survival of a seabird. *Ecology*, 99(5), pp.1063–1072.

Chavez, F.P., 2003. From anchovies to sardines and back: Multidecadal change in the Pacific ocean. *Science*, 299, pp.217–221.

Conningarth Economists, 2014. *A Manual for Cost Benefit Analysis in South Africa with Special Reference to Water Resource Development*. WRC Report GT177/02. Pretoria: Water Research Commission.

Conrad, J.M., 2019. Dynamic optimization, natural capital, and ecosystem services. In H. Kaper and F. Roberts (Eds), *Mathematics of Planet Earth*, Vol. 5. New York: Springer, Cham.

Costanza, R., 1999. The ecological, economic, and social importance of the oceans. *Ecological Economics*, 31(2), pp.199–213.

Crookes, D.J. and Blignaut, J.N., 2019. An approach to determine the extinction risk of exploited populations. *Journal for Nature Conservation*, 52, p.125750. doi:10.1016/ j.jnc.2019.125750

DAFF, 2017. *Initial Recommendation of the Small Pelagic Scientific Working Group for the Sustainable Management of Small Pelagic Resources for the Season 2018*. Fisheries/2017/dec/swg-pel/39. Pretoria: Department of Agriculture, Forestry and Fisheries.

DEA, 2019. Draft 2nd biodiversity management plan for the African penguin (*Spheniscus demersus*). *Government Gazette*, 52(42775). Pretoria: Government Printers.

De Wit, M.P., Crookes, D.J., Blignaut, J.N., de Beer, Z.W., Paap, T., Roets, F., van der Merwe, C., and Richardson, D.M., 2021. Invasion of the Polyphagous Shot Hole Borer beetle in South Africa: A preliminary assessment of the economic impacts.

Ford, A., 2009. *Modeling the Environment* (2nd edn). Washington, DC: Island Press.

Hutchings, K., Clark, B.M., and Turpie, J.K., 2014. *Assessment of the Socio-economic Implications of a Reduced Minimum Sardine TAC for the Small Pelagics Purse-seine Fishery*. Final report submitted to SAPFIA.

Isaacs, M., 2016. The humble sardine (small pelagics): Fish as food or fodder. *Agriculture & Food Security*, 5(1), pp.1–14.

Lee, M.-Y., Steinback, S., and Wallmo K., 2017. Applying a bioeconomic model to recreational fisheries management: Groundfish in the Northeast United States. *Marine Resource Economics*, 32(2), pp.191–216.

Lewis, S.E.F., Turpie, J.K., and Ryan, P.G., 2012. Are African penguins worth saving? The ecotourism value of the Boulders Beach colony. *African Journal of Marine Science*, 34(4), pp.497–504.

Lindegren, M., Checkley, D.M., Rouyer, T., MacCall, A.D., and Stenseth, N.C., 2013. Climate, fishing, and fluctuations of sardine and anchovy in the California current. *Proceedings of the National Academy of Sciences*, 110, pp.13672–13677.

Lo, N.C.-H., Smith, P.E., and Butler, J.L., 1995. Population growth of northern anchovy and Pacific sardine using stage-specific matrix models. *Marine Ecology Progress Series*, 127, pp.15–26.

Petersen, S.L., Ryan, P.G., and Gremillet, D., 2006. Is food availability limiting African Penguins Spheniscus demersus at Boulders? A comparison of foraging effort at mainland and island colonies. *Ibis*, 148(1), pp.14–26.

Robinson, W.M.L., Butterworth, D.S., and Plagányi, É.E., 2015. Quantifying the projected impact of the South African sardine fishery on the Robben Island penguin colony. *ICES Journal of Marine Science*, 72(6), pp.1822–1833. doi:10.1093/icesjms/fsv035.

Saayman, A., Saayman, M., and Naudé, W.A., 2000. The impact of tourist spending in South Africa: Spatial implications. *South African Journal of Economic and Management Sciences*, 3(3), pp.369–386.

Van Zyl, H.W. and Kinghorn, J.W., 2018. *The Economic Value and Contribution of the Simon's Town Penguin Colony. Report to the City of Cape Town*. Cape Town: Independent Economic Researchers.

Waller, L., 2011. *The African Penguin Spheniscus Demersus: Conservation and Management Issues*. PhD thesis, University of Cape Town, Cape Town.

Weller, F., Cecchini, L.A., Shannon, L., Sherley, R.B., Crawford, R.J., Altwegg, R., Scott, L., Stewart, T., and Jarre, A., 2014. A system dynamics approach to modelling multiple drivers of the African penguin population on Robben Island, South Africa. *Ecological Modelling*, 277, pp.38–56.

8 Discussion and conclusions

This book highlights some of the uses of co-evolutionary models as a means of providing strategic (long-term) management of environmental change (five to ten years). These models may be used for a variety of purposes towards this aim, for example in order to estimate unknown biological and economic parameters; understanding the underlying behaviour of a system; forecasting future values of stock variables; understanding trade-offs between environmental, social and economic objectives; and how to facilitate win–win or double and triple bottom-line outcomes.

Chapter 2 presented the different co-evolutionary models and when they can be used. Then, in Chapter 3, the simulation modelling tool (Vensim) was proposed, and steps in the modelling process presented, along with validation, which is crucial when curve fitting is required. Chapter 4 showed how to use numerical methods to estimate unknown parameter values by curve fitting in Vensim (using Markov chain Monte Carlo simulation). Once the model is fitted and validated, and a good fit is achieved, it may then be used for strategic purposes.

Chapter 5 gave an example from the rhino poaching literature to show how these models could be used for curve fitting and forecasting, and also for understanding the drivers of poaching. The best model provided a good forecast of rhino population dynamics (based on the mean absolute percentage error statistic, from 2013 to 2019, over seven years). This demonstrates that these models are well suited for long-term (strategic) advice (five to ten years at least).

Chapter 6 showed the use of these co-evolutionary models for understanding when co-operation may be desirable. A two-player game was proposed based on the prisoner's dilemma game. The two players could be two species or another environmental variable, it could be an environmental and economic variable, or two economic variables. In some cases, social trade-offs could also be modelled. The analysis showed that while mutualism was optimal under resource abundance, under resource scarcity it may not be the

DOI: 10.4324/9781003247982-8

optimal outcome. There is an incentive to defect. Management may therefore need to recognise that mutualism may sometimes be suboptimal, but may nonetheless desire to force this outcome in order to achieve win–win outcomes between different sectors or species.

Chapter 7 gave the example of the African penguin to highlight conditions when co-operation may be suboptimal. The model gave a reasonable fit with the historical data (in terms of the mean absolute percentage error statistic, 1987–2017) and is therefore better suited for long-term strategic advice rather than short-term tactical advice. Closing the sardine fishery has win–win outcomes for the environment (penguin and sardine populations) and the economy (tourism values). However, there is a social externality in that the poor are disadvantaged. The government may therefore need to make trade-offs over the long term and accept a suboptimal outcome in order for there to be a triple bottom line (or three sectors or species benefitting rather than just two).

Taking Chapters 6 and 7 together, under resource scarcity, it is optimal for one sector or species to benefit and all others to lose. However, two sectors can benefit equally, and produce results that are in excess of maximum sustainable yield, or three sectors can benefit, in which the individual benefit declines but three sectors or species benefit. In the penguin model of Chapter 7, for three sectors to benefit, sardine's steady state was at approximately $0.5K$ (which is maximum sustainable yield), penguins were at approximately $0.07K$ (which is still very low, but nonetheless stable), and no social externality. For two sectors to benefit, sardines go to carrying capacity, penguins go to $0.24K$ but there is a social externality. The final decision in making these trade-offs, and determining which sectors or species ultimately benefit, lies with the government. We can therefore term this the government's dilemma.

Which strategic model is best? This will depend on the type of model. In the case of bioeconomic models, when it comes to curve fitting, the Pella and Tomlinson model, along with the Schaefer production function, seems to be a good place to start. If this doesn't work, try a different catch function or an adaptive expectations hypothesis. In the case of ecological models, a much wider range of models may be appropriate and will need to take into consideration the nature of the interactions between the species. Ultimately, the modeller will need to be led by the literature, expert opinion, and the goodness of fit the model achieves with the historical data in order to inform the best approach. An iterative process may therefore be required before a model is found that adequately fits that data.

When is a model fit good enough to use for strategic advice? As we saw in Chapter 7, it is not necessary for a model that is used for long-term (strategic) advice to provide the fine-scale accuracy that a tactical (three- to five-year) model would need to provide. I would say, any model with a mean absolute percentage error (MAPE) of around 50% or less (in other words a reasonable fit or better) is probably suitable to use for strategic advice. However, it will depend on the exact use of the model and the quality of the data. If the data are highly variable, in some cases a poorer fit may be acceptable, provided the model captures the long-term trend adequately. Further evaluation of such models will be necessary to ascertain the suitability of these models for strategic advice.

Index

Printed in the United States
by Baker & Taylor Publisher Services